Claude Martin

The Rainforests of West Africa

Ecology – Threats – Conservation

English Translation by Linda Tsardakas

Birkhäuser Verlag
Basel · Boston · Berlin

Library of Congress Cataloging-in-Publication Data

Martin, Claude:
[Regenwälder Westafrikas. English]
The Rainforests of West Africa : ecology – threats – conserva-
tion / Claude Martin ; English translation by Linda Tsardakas.
p.   cm.
Translated from the German.
Includes bibliographical references and index.
ISBN 0-8176-2380-9 (U.S.)
1. Rain forests–Africa, West. 2. Rain forest ecology–Africa,
West. 3. Forests and forestry–Africa, West. 4. Forest conserva-
tion–Africa, West. I. Title.
QH195.W46M3713   1991
333.75'0966–dc20

Deutsche Bibliothek Cataloging-in-Publication Data

Martin, Claude:
The Rainforests of Westafrica : ecology – threats – conserva-
tion / Claude Martin. English transl. by Linda Tsardakas. –
Basel ; Boston ; Berlin : Birkhäuser, 1991
Dt. Ausg. u. d. T.: Martin, Claude: Die Regenwälder Westafrikas
ISBN 3-7643-2380-9 (Basel . . .) Gb.
ISBN 0-8176-2380-9 (Boston) Gb.

© 1991 Birkhäuser Verlag, 4010 Basel, Switzerland
Book Design: Albert Gomm swb/asg, Basel
Printed in Germany
ISBN 0-8176-2380-9
ISBN 3-7643-2380-9

Preface

Nowhere else in the world did industrialized countries leave such early marks in the rainforest as in West Africa. Past and present developments here are in one way or the other significant for rainforests on other continents as well. West Africa is a pioneer in both a good and a bad sense. This is reason enough to take a closer look at the history of moist tropical West Africa.

Until recently, no one really seemed to be interested in the rainforests except for a few specialists. The world's scientific community neglected to study the incalculable riches of tropical forests, to make the public aware of them and their due importance. Although interdisciplinary research has been a popular topic for some decades now, it was not applied to just the most complex habitat on earth. Scientists from all fields studied only that which was easiest to record, seemingly blind to a myriad of details awaiting closer examination. Botanists went about establishing their herbariums and paid much too little attention to the vegetation as a whole, or to the significance of useful plants for local populations. Zoologists, too, busied themselves with collecting and describing species. Anthropologists, on the other hand, tended to overlook faunal details: in their ignorance of the animal world, they wrote of tigers and deer in Africa. And finally, foresters saw neither the forest nor the trees for the timber – and even confused rainforests with monocultures of fir trees. They overlooked the significance of collectable forest products and wildlife as they still do today.

The destruction of tropical rainforests is not a problem due only to a lust for profit and land. It is just as much the consequence of a scientific community's neglect. Scientists have not yet even set priorities for research, let alone developed methods capable of functioning in the canopies of the rainforest. The industrialized world should therefore not be surprised at the fate of tropical rainforests. We have done little to conserve them and all the more to destroy them. And it is the forest people who have suffered along the way, their needs have been entirely ignored. Almost nowhere is their culture given the respect it deserves.

In this book, I have dared to attempt an integral presentation of West African rainforests: I have gathered old and new knowledge – from biology to forest utilization and the cultural background to forest conservation. And in spite of the years and days I personally spent in these forests, I am aware that I can give only a partial picture. Nevertheless, if this book can convey a feeling for and an understanding of the rainforest as a whole, if it can contribute to the conservation of these forests, then it has fulfilled its purpose.

Summer 1990

C. Martin

Salvation or sacrifice for the rainforest? As in other West African countries, forest reserves were established in Ghana early in this century (green areas). Forest-dwelling people were largely deprived of their traditional rights to land-use. The reserves were henceforth meant to ensure the production of tropical timber.

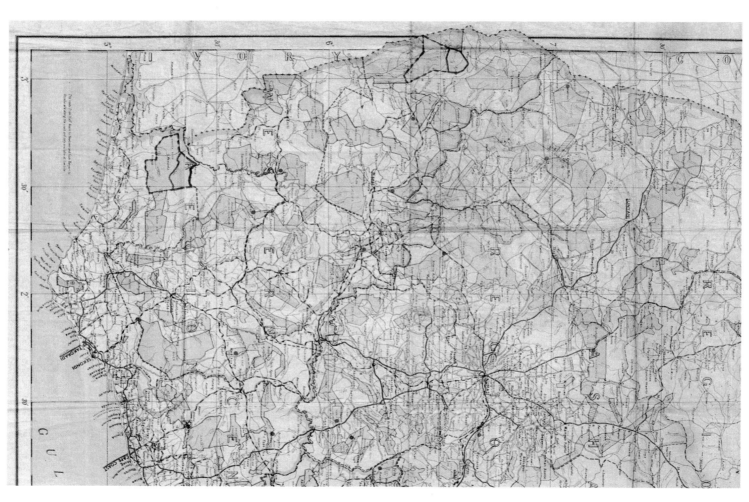

Contents

Immigrant farmers use axes and cutlasses to clear what remains of the forest once timber exploiters have moved on. The dried remains of the cut vegetation are then set afire. But slash-and-burn clearing affords only a few harvests of corn, plantains and cassava. This kind of forest destruction is mostly limited to areas previously opened up by timber exploitation.

Europe Casts Its Spell upon the Rainforest

∨ Rituals must be held to sooth the invisible spirits of the forest: Sacrificial ceremony in Debiso, Ghana.

The thunder of drumbeats can be heard far and near, the smell of roasting meat permeates the air, palm wine and schnaps are passed around. Why has the Chief called to celebrate the yam festival? Nobody really knows. Nevertheless, drumbeats spread the news throughout the forest. With mounting excitement, the ornamental stools are removed from storage. The people of Debiso in Western Ghana seldom actually see their ancestral inheritance. The chairs are usually kept by the village chief, locked away in his personal storeroom. With the aid of his medicine men, the Chief is capable of speaking with the spirits of the forest. During three days of festivities, ceremonies will be held at specially chosen sites outside the village to appease the invisible beings.

The rainforest rises like a great fortress behind the small village. In the clearing, simple tin-roofed huts stand unsheltered in the scorching sun. Even the chickens appear to be the smallest of birds against the backdrop of the rainforest. Few villagers have ever seen the end of the forest although the "Trotro" – an old Bedford truck – makes daily trips on the timber transport route from Kumasi to the border of the Côte d'Ivoire*.

* In 1985, the official name of the Republic of the Ivory Coast was changed to "Côte d'Ivoire" in all languages.

The oil palm (*Elaeis guineensis*) is native to rainforests in West and Central Africa. In 1848, the Dutch introduced the species to Java.

Until a few years ago, the village could only be reached on foot via winding paths through the deep forest.

The Chief wears a heavy gold-plated crown for the ceremony. The same ponderous headdress had been born by his forefathers in leading the Sefwi tribe into battle against the Ashantis in brutal jungle wars of the past. That was before the hunting camp which once stood at this site had evolved into a permanent settlement. Only the name "Debiso" remains to remind one of the place where two forest elephants killed by Sefwi hunters came to fall upon each other. Sacrifices are made to the forest and its spirits every three years at the Yam Festival. But the dwarfs must

be appeased more often. Many West African tribes believe strongly in the lore of the little people. The dwarfs, invisible creatures who live at secret places in the forest, must be given due respect. Otherwise, their playful pranks may easily become evil tricks. They have even been known to kill on occasion.

Natural clearings are rare in the deep forest, occurring only where the granite base of the African continent rises to the surface. The Sefwis call the clearings "Apaso" – the dwarfs' gardens. The "gardens" are magical places where sacrifices are made with great care according to the wishes of the village chief and his medicine men. It is also they who determine which of the village elders may accompany them along the twisting forest path leading to the secret site.

A stagnant pool of greenish water awaits the visitors. According to tradition they do not speak and have entered from one direction only. The medicine man now summons the dwarfs in a loud voice and invites them to hear his humble words. After a long sermon, the sheep which has been brought from the village is sacrificed – its throat slit allowing its blood to flow upon the objects spread upon the bare rock. Bottles of schnaps are poured onto the ground as an offering to the dwarfs, no less is drunken by the Chief and his companions themselves for they are fearful and alcohol relieves their fear. It is said in Debiso that some men have never returned from the clearing because they were unable to appease the "little people". The dwarfs are the keepers of the forest and of the animals and do not hesitate to punish wrongdoers. The people strongly believe in this tradition and take care not to anger the little sentries. Thus, the responsibility of protecting the forest so vital to the Sefwi tribe is in the hands of these small creatures. But the dwarfs never reckoned with the Europeans!

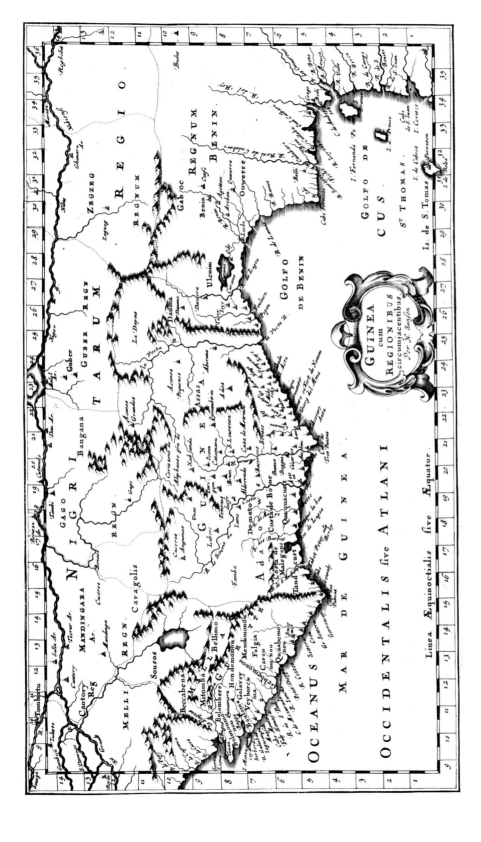

This map of West Africa dates back to 1679 and shows several European trade settlements along the coast. The interior of the continent, however, remained a mystery to the Europeans except for what they heard from Arab and native traders.

A Source of Raw Materials for 500 Years

The sea-route from the major harbors of southern England, Germany or Holland to the coast of the Gulf of Guinea extends over 600 kilometers. Neither the rainforests in Latin America nor the forests of Southeast Asia are closer. West Africa's proximity to Europe has influenced its trade relations for centuries and left its mark on the rainforest.

Europe's first trade relations with the West African coast date back to the 15th century. The first trade agencies consisted of nothing more than storage houses and fortifications along the coast. A small strip of land sufficed to develop trade from the coast famed for its rich natural resources. The rest of the African continent was still a mystery to the Europeans. Nevertheless, trade boomed: gold, ivory, cola nuts and slaves were shipped from the coast of western Ghana in 1700. British, Dutch and Danish trading posts vied for the best ports. Not every trade settlement had been founded at an ideal location and the sailors had to battle against rough waters [1].

The Europeans did business with Arab slave dealers from the north and certain tribes of rainforest people who delivered goods and slaves from the interior of the continent to the coast.

Early oil palm plantation in Nigeria: The light red fruit grows at the base of the leaves above a squat trunk which can, however, reach a height of 15 m. Today, Asia is the major producer of palm oil where the species is cultivated on immense plantations. Palm oil covers approximately one seventh of the production of vegetable oil worldwide.

Slavery was not a European custom alone. The Ashanti tribe, for example, kept slaves for their own use. Entire villages in the savannah north of the West African rainforests were taken into bondage if they did not succeed in fleeing their captors. This led to a depopulation of large areas which still remain thinly populated today [2]. The poor uprooted people from the north were considered nothing more than a good to be traded – even by the West Africans. Slaves were a cheap commodity in contrast to salt which was quite expensive. It was imported from the north in caravans and traded like gold. In the interior of Benin, Togo and in the Volta region of Ghana, a handful of salt was known to be worth one to two slaves [1].

The Danes were more liberal than other Europeans active in West Africa at the time, abolishing slavery in 1802. The British Government followed suit five years later with a ban on slave export which they applied to a number of other trading countries as well. The British confiscated foreign slave ships and freed their unfortunate passengers. The Europeans' new respect for other races, however, posed a problem for some forest peoples; the Ashantis were left with no takers for their wares and opted to settle their captives as planters. The slave trade was replaced by the export of palm oil [1]. Inadvertently, the British were therefore actually responsible for creating export-oriented agriculture in West Africa. In 1850, they began to expand their influence.

The Beginning of Commercial Exploitation

In 1879, following 400 years of trade with the West African coast, the Europeans only ruled over large segments of the population in French-Senegal and on the British "Gold Coast" (Ghana). Except in Senegal, European administrations never controlled land more than a few miles inland. The British colonies of Gambia, Sierra Leone and Lagos were nothing more than small enclaves in a region still largely under African rule [3]. Within but a few years, however, the situation was to change drastically. In 1870, gum copal from trees of the genus *Daniellia* found in closed rainforests was being exported for the manufacture of varnish in increasing quantities. The European demand for rubber tapped from *Funtumia* trees also rose steadily after 1883. Following a number of unsuccessful attempts, the export of agricultural products rapidly increased. The first successful oil palm plantations in southern Ghana exported up to 30000 tons of oil by

Cocoa beans drying in the hot sun in Adjufua, a forest village in Ghana. Until a few years ago, Ghana was the world's largest cocoa exporter.

1884. In 1895, Ghana also exported 63 tons of coffee [1].

During the last quarter of the 19th century, rainforest in Ghana and Nigeria increasingly gave way to cocoa plantations. In contrast, cocoa was not brought to the Côte d'Ivoire until 1912 when it was introduced from Ghana. Lowland rainforests in the southern part of the rainforest belt suffered most from the cocoa boom. Cocoa cultivation was especially successful in the rural areas of Ghana near Accra and in the forest region near Kumasi. In 1911, after only 26 years of profitable production, Ghana rose to become

the top exporter worldwide and managed to hold its position for a number of decades while continuingly increasing production.

First attempts at logging in Ghana had been undertaken near Axim in 1800. They were abandoned, however, due to technical difficulties. It was not until 1887 that logging operations were again established on an "experimental basis" [1]. The British, who now ruled the larger part of southern Ghana, eliminated the political barriers to allow timber to be floated freely to the coast on the major rivers Tano, Ankobra and Pra. The "experiment" was to have serious con-

Commercial timber plantations in Nigeria were established early under colonial rule. This photo taken in 1911 portrays the British Conservator of Forest, H.N. Thompson, on a teak plantation in southern Nigeria.

∨
Today, most of West African cocoa is still produced on small-scale farms. The yellow cocoa fruits are broken open and the slimy pulp with the seeds (cocoa beans) is spread to ferment for a few days on banana leaves. Only then can the beans be washed and dried.

sequences: Seven years later, the export of tropical timber had reached 450000 cubic feet (approx. 285 m³) and the colonial government had begun to distribute timber concessions to European investors. By the year 1913, Ghana's exports had gradually risen to three million cubic feet (approx. 85000 m³). World War I, however, brought difficult times and the export volume fell back to a much lower level. Nigeria was exporting similar quantities of tropical timber and early in the 19th century the country established its own forest service. Only African mahogany (Khaya ivorensis) and smaller quantities of sapele (Entandrophragma cylindricum)* were actually in demand then for European furniture production.

Until World War I, the British largely controlled the export of tropical timber from the Côte d'Ivoire as well. Mahogany was felled near the Bia River and the logs were floated to the Abi Lagoon near Assinie. The consequent depletion of the forest along the shore led to trees being felled further inside the forest. The logs were hauled to the nearest rivers along corduroy roads [4]. In time, other African hardwoods such as makore and iroko gained popularity in Europe. Nevertheless, export statistics showed African mahogany to be at the top of the list until 1951. But until after World War II, timber exports from West Africa showed slow growth. The fluctuating market demand was not the only reason. Transporting logs from distant locations to the coast posed nearly insurmountable technical problems. The interior rainforests were thus spared from large-scale exploitation until the 1950s when the mechanization of forest operations changed the situation dramatically. Bulldozers were brought in to push their way through the forest and trucks were able to haul entire trunks from deep within the forest to the coast.

The European demand for tropical timber increased steadily throughout the 1950s. In the Côte d'Ivoire, roundwood production sky-rocketed from 400000 cubic meters in 1958 to more than 5000000 cubic meters in the 1970s [4]. Nowhere along the Gulf of Guinea did the exploitation of the rainforest proceed at such a fast pace as in the Côte d'Ivoire. Neighboring Liberia was not similarly affected until the mid-1960s. While the export of timber from the West African coast drastically increased after World

* The author will henceforth refrain from listing the scientific names of commonly exported species except in special cases. Annex 1 lists both the scientific and trade names.

War II, the export of other forest products came to a near standstill. Artificial resin had long since replaced natural gum copal in the production of varnish and the rubber grown in West Africa today is actually caoutchouc from Brazilian Hevea rubber trees grown on large plantations. Just as for the cultivation of cocoa, wide areas of rainforest were also cleared for Hevea and oil palm plantations.

Early Attempts at Conservation: Theory

It would be unfair to assume that the Europeans were interested only in exploiting the natural resources of their African colonies with no consideration for their future. Early in the history of commercial timber exploitation, the colonial government in Ghana adopted the Timber Protection Ordinance of 1907 which banned felling commercial species of a lesser diameter [1]. In Togo, a German protectorate from 1884–1919, forest conservation had been a concern from the start of colonial rule. The country was endowed with a far smaller area of rainforest than other West African nations. In 1907, just as forestry regulations were adopted in Ghana, a conference was held in Berlin on the reforestation of Togo. The Germans decided to invest mainly in teak, an exotic species well-adapted to the drier climate. In addition, indige-

nous hardwoods were also chosen: doussie, sasswood, African mahogany and iroko, the kapok-tree (*Ceiba pentandra*) and "chew-stick" (*Anogeissus leiocarpus*). The bark and leaves of "chew-stick" are of medicinal value, only the roots are used as chewing sticks. Thirteen million trees are said to have been planted in Togo early this century [5].

In Nigeria, the British had begun with reforestation in the Benin District even before the forestry administration was established. Tens of thousands of African mahogany trees were planted annually. Between 1901 and 1910, trees of several species native to Nigeria were planted: iroko, obeche, limba and mahogany. A few exotic species were also introduced: teak, cedrela and eucalyptus. In 1920, the Senior Conservator of Forests of Nigeria, A. H. Unwin stated, "Despite many failures owing to experiments on bad soil and seasons of extreme drought, the growth of the trees gives the greatest promise of mature trees, or at any rate merchantable trees, being grown in a comparatively short period" [5].

Soon after, the British realized that protecting the forest as a whole was even more important than planting individual trees. With the establishment of the forest service, the administration began to demarcate forest reserves, which were henceforth to provide a source for permanent and controlled timber exploitation. In order to preserve the steady flow of rivers to lower lying regions, watershed areas were especially given reserve status to protect against flooding, erosion and drought. Shelterbelt forest reserves bordering on savannah territory marking the transition to the Sahel were established to keep back the hot desert winds from the north. In Ghana, the Forest Ordinance of 1911 allowed the colonial governor to give all uninhabited forest territory forest reserve status, a measure which stripped the surrounding inhabitants of their traditional user rights. The people pro-

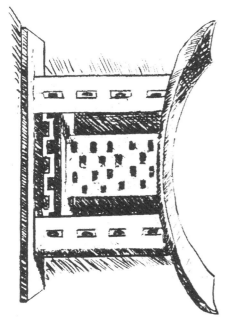

The Ashanti "stool" is a symbol for ancestral land.

The "Nana" or village chief still wears a gold-plated crown and a traditional robe of Kente cloth for ritual ceremonies. But his rights have been severely limited through the centralization of the forest administration.

tested strongly since their livelihood depended on just that which was thus prohibited by the ordinance, namely shifting cultivation and gathering forest products such as essences, fibers, fruits and nuts. The Aborigines' Rights Protection Society lodged a complaint with the Governor for disrespecting the traditional land tenure system [1].

Reality: Forest Inhabitants Resist

West African tradition holds land to be common property belonging to one or more communi-

ties. The members of the community are entitled to use as much land as they need to clear and cultivate. Should land no longer be needed, it falls back to the community. Sale of property is not allowed since it does not belong solely to the living users but also to their ancestors buried on the land and to future generations. For Akan tribes in the eastern Côte d'Ivoire and in Ghana, the earthly resting place of their forefathers' souls is symbolized by the "stools". The stools, each cut from a single block of wood and richly embellished with ornamental carvings, represent the common property, also termed "stool

land". During rituals held at regular intervals, offerings of food are placed about the stools. Sacred wine and schnaps are used to douse the precious inheritance. The British were well aware of the importance of the stools in regard to property rights. In 1900, upon the conquest of the mightiest Akan group, the Ashanti tribe, British governor Sir Frederic Hodgson demanded possession of the "Golden Stool". In a storm of protest led by the mother of their imprisoned King Kwaku Dua III, the Ashantis attacked the British fortress in Kumasi. The remembrance of that event probably played a role in their reluctance to strictly enforce the new forest regulations. Local chiefs were still quite influential. The legislation was in fact temporarily withdrawn until the newly established forest service published long lists of forest reserves between 1922 and 1926. As one may expect, the forest reservation program was not at all popular with the local inhabitants, who felt they were being denied their traditional rights. By the year 1939, an area of 14800 square kilometers of Ghanaian rainforest had been declared forest reserves, which totalled 19% of the country's rainforest cover [1].

In 1926, the French colonial power began similar attempts at forest conservation in the Côte d'Ivoire. By 1956, 43000 square kilometers of forest had been declared protected areas (forêts classées). The local reaction was no more favorable than in Ghana. Many of the Ivorian forest reserves were so heavily damaged by illegal slash-and-burn clearing that redefinement was necessary in 1966. Unfortunately, the problem was not easy to solve. Illegal farming on reserved land continued largely uncontrolled [4].

The Consequences of Centralization

Throughout the century, West African peoples have increasingly lost authority over their vast

areas of forest. Colonial regulations withdrawing their self-responsibility within the newly established forest reserves were not the only reason. Logging activities made the forest more accessible. Transport routes, although crude, not only allowed timber to be carried out of the forest, but also let new settlers come in. Immigrant farmers came from other parts of Africa, especially from the north where the dry climate forced many to look elsewhere for their livelihood. The new settlers, however, were not familiar with the rainforest. In contrast to West African forest peoples who had lived and worked for generations in the forest, they did not know how to deal with the sensitive soil. Traditional West African forest farmers never do long-term damage to the land. The area surrounding a village is cultivated on a rotational basis and only small areas are cleared and planted for a few years before the next ones are cleared. Once a plot of land has been depleted of its nutrients, it is left under fallow for natural regeneration to occur and the rotation continues around the village allowing several years to pass before the same plot is replanted. This form of shifting cultivation is still practiced today in less populated areas in the western Côte d'Ivoire, in parts of Liberia and in western Cameroon.

The opening up of the forest for logging activities and increased population led to the collapse of traditional land tenure and land-use systems in most West African forests. Coffee and cocoa plantations contributed in no lesser sense to the disintegration of the traditional systems. Although land is considered to be community property, many tribes believe that which grows on the land to be the sole property of the planter. Coffee trees, cocoa trees and oil palm trees thus belong to the farmer and not to the community. Should the community wish to reclaim its land, the farmers must be compensated for their fruit trees [2]. Many villages therefore do not allow immigrant farmers to plant fruit trees

< An old Bedford truck has been remodelled for public transportation along a logging road in western Ghana up to the border of the Côte d'Ivoire. Until a few years ago, only a footpath led through this part of the forest. Now immigrant farmers have moved in.

Table 1
The status of closed broadleaved forests ("rainforests") in West Africa from Sierra Leone to Nigeria in 1985. Estimates by FAO/UNEP [6].

Status of forest	km²	%
Undisturbed, productive	21 260	4.1
Undisturbed, unproductive	64460	12.5
Logged	45870	8.9
Managed	11670	2.3
Forest fallow	370820	72.1
Total West Africa	514080	100.0

Table 2
Annual decrease in area of closed broadleaved forests ("rainforests") from 1981 to 1985. Estimates by FAO/UNEP [6].

Forest Type	km²	%
Untouched, productive	210	2.9
Untouched, unproductive	340	4.7
Logged	6650	92.4
Total West Africa	7200	100.0

Note: The FAO statistics use the term "productive" to designate commercially exploitable forests. The term "unproductive" refers to protected areas and inaccessible terrain, for example hilly forest areas. "Logged" refers to forests subjected to selective exploitation and "managed" refers exclusively to figures for Ghana where most forest reserves are considered "managed" regardless of the actual situation.

on the land for which they have received cultivation rights. The result is that the settlers then plant only corn, plantain and cassava, all of which quickly deplete the soil of its nutrients. After only a few years, the planters find themselves forced to move on to a new plot of land. Those who are determined to remain at one location for a longer period usually choose to plant coffee and cocoa. They do their best to gain possession of land against the traditional rules. These circumstances have led to a wider distribution of plantations and villages throughout the forests of West Africa. West African tribes thus had good reason to

resist the centralization of the forest administration at the turn of the century. They tried to defend their culture and their ancestors – without success. But neither did the West African forest services achieve their goals. Forest legislation based on European principles was inadequate in the light of African reality. In trying to encourage the sustainable development of the rainforest, the forest services lost control over the consequences of their ordinances. Even today, West African forestry departments lack the necessary foresight, control and overview to deal effectively with the situation. Laws and regulations aimed at protecting the forest often lead to just the opposite.

Large-scale Exploitation and the Consequences

It was not until the 1970s that the inadequate implementation of forest regulations became apparent. Following World War II, commercial exploitation had increased to such an extent that no West African forestry department was capable of enforcing the law. In comparison with rainforests in other parts of the world in 1973, Africa showed the greatest area encroached upon by logging activities although African timber production measured only one third that of Asia. This fact signalized very extensive exploitation practices in African rainforests which were continually opening up new areas.

In 1985, the U.N. Food and Agricultural Organization (FAO) estimated 72% of West African rainforests to be fallow land: destroyed forest, plantations or secondary bush (Tab. 1). Between 1981 and 1985, the remaining areas of undisturbed forest on the Gulf of Guinea were being opened up at the rate of 1640 square kilometers annually [6]. Many timber companies adopted highly selective logging practices to meet specific demands or merely for reasons of profit. Only

The tropical timber boom is past in Axim, an impoverished town on the coast of Ghana. The humid climate has eaten away at the mansion of the one-time timber magnate George Grant. Now a fig tree has taken possession of the palace from within.

the best specimens of mahogany, utile, sapele and makore were extracted from the forest. With few exceptions, however, such selective timber exploitation foresaw a one-time use only and did not aim at long-term sustainable use [7]. During the past decade, each year has seen an average loss of 7200 square kilometers of West African forest. The major cause is burning and clearing by migrant farmers. But over 90% of the areas destroyed were forests previously opened up by timber companies (Tab. 2). Logging roads not only serve to carry timber out of the forest but also pave the way for settlers to come in. What timber companies leave standing is devastated by slash-and-burn farming. Unfortunately, the resulting increase in agricultural area is minimal: since the soil quality of the land thus gained affords only a few harvests, the process of burning and clearing continues. The planters follow the timber exploiters further into the rainforest leaving unproductive land behind.
Today, we look back upon a half a millenium of trade relations with Europe – relations which were rarely in favor of West Africa. European demand dictated West Africa's export of natural resources. During times of war or upon the discovery of an alternative product, Europeans lost interest in trade. Whether dealing in slaves or hardwood, Africa's fate was typical for a deliverer of raw materials. Unlike the eastern and southern parts of the continent at higher altitudes, West Africa did not attract many white settlers, its climate being unpleasantly hot and humid. The whites were interested only in what could be exported. West Africa became Europe's major source of raw materials and remained a loyal trader even in less profitable times. But West Africa was poorly paid for her loyalty. What began as a sustainable use of forest products developed into a timber export economy nothing short of all out exploitation. In the words of tropical forest expert Professor H. Steinlin, "The exploitation can be compared to that of a mine, a natural resource is being excavated without guarantee of a sustainable production" [8]. This attitude has had serious consequences for the forest and last but not least for the local people.

∨
The concessionaire found no market for this and other logs from his land behind Axim. The recession of 1975 brought hard times for the timber industry. All that remains here is the mark "GG" to remind us of George Grant at the height of his success.

The Distribution of West African Rainforests

Until last century, European travellers to Africa believed the continent to be covered with rainforest from just about one end to the other. Sailors along the coast of West Africa between latitudes 10°N and 10°S, from Sierra Leone to north of the Congo River, observed nothing but seemingly impenetrable jungle. Trees lined the coast just short of the water's edge and damp foggy mists hopelessly hid the interior of the forest from inquiring eyes. A single stretch of savannah 300 kilometers long running from Accra in Ghana to Cotonou in Benin attested to the fact that rainforest was not the only form of African vegetation. This arid zone is known today as the Dahomey Gap. Early pioneers most likely believed it to be backed by dense jungle growth. After all, Africa looked no different when approached from the Indian Ocean. With wet coastal forests lining the eastern side as well, the continent appeared to be absolutely impenetrable and quite threatening. Even as the Europeans began to press their way inland during the 19th century, their impression of endless jungle was not to change. Most expeditions made their way along rivers lined with gallery forests, a common characteristic of waterways in West Africa. Although open savannah was but a few kilometers beyond the river's edge, travellers believed themselves moving through vast expanses of wet jungle.

Most of Africa's rainforests are lowland forests: Sunset over the border area between Ghana and the Côte d'Ivoire.

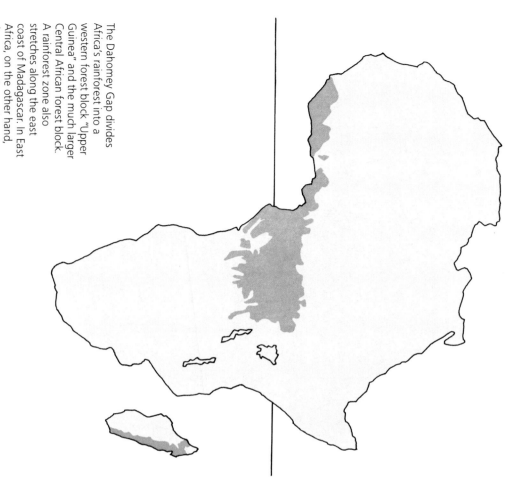

The Dahomey Gap divides Africa's rainforest into a western forest block "Upper Guinea" and the much larger Central African forest block. A rainforest zone also stretches along the east coast of Madagascar. In East Africa, on the other hand, there are only small patches of rainforest. Mountain forest types are not shown here.

While European knowledge of African geography was largely based on assumptions and at best vague descriptions until far into the 19th century, the Arabs were at least able to describe West Africa north of the rainforests. They had dealt for centuries with Hausas near Lake Chad and with the Mande traders from the area of Bamako in former Mali. The traders brought small amounts of gold dust and ivory with them from the rainforests on the west coast. Cola nuts were popular with north African moslems because of their stimulating effect. The Arabs traded salt, glass pearls and textiles in return for the luxury goods from the coast [3]. So the forest-dwelling people of West Africa had traded across the Sahara with the Mediterranean region at least since the late Middle Ages. The Arabs had been familiar with the area long before the Europeans began to officially discover the continent beyond the coasts last century. International trade was not new to Africa. In spite of beliefs held up until a relatively short time ago, there are no indications that the distribution of rainforests over the past millenia was notably greater than today's area of distribution and transitional areas. On the contrary: During the European ice ages, Africa's climate is assumed to have been cooler and drier resulting in the rainforests shrinking to small isolated refuges. The consequences of this intermittent shrinking process can be found today in the distribution of certain plant and animal species. So for thousands of years rainforests have covered only a part of the African equatorial zone and have never accounted for much more than 10% of the continent's total area.

What is Rainforest?

The numerous classification systems used for African vegetation zones seem to be a jungle of their own. Until the 1950s, classifications were made by botanists and geographers based only

The Sassandra River in the western Côte d'Ivoire forms a zoogeographical barrier for numerous animal species which occur only in Liberia and in the Taï area of the western Côte d'Ivoire. The Sassandra originates in the savannah north of the rain-forest zone and there, like many African rivers, it is lined by moist gallery forests.

on the knowledge of one specific area or country. This resulted in a confusing multitude of terms for comparable types of vegetation as well as a number of identical terms for very different types. The fact that such vocabulary existed in several languages complicated matters still further. African savannah vegetation gave rise to particular confusion since the transition to woody plant cover is gradual and cannot be adequately described by terms like "tree savannah" or "thornbush savannah". Translations of the English words "scrub" and "bush" led to further misunderstandings among scientists

in continental Europe. Similarly, the term "steppe" used to depict the driest forms of savannah vegetation is also inexact. This general confusion led to an overall avoidance of the term savannah among the experts. More recent scientific literature tends to distinguish between "grassland" and "woodland" or "bushland" of varying density and composition. An additional problem is posed by the classification of secondary savannah forms which have evolved as the result of human influence as well as fires in woody areas and in forests. The increasingly rapid transformation of natural landscapes has only added to an already confusing terminology.

Today, however, there are much better aides to classifying African vegetation. In 1956, at the suggestion of the "Scientific Council of Africa South of the Sahara" [9], specialists in plant geography met for the first time in Yangambi (Zaïre) to found a uniform system of nomenclature. The "Yangambi classification" led to the first adequate vegetation map of Africa published in 1959 and edited by the "Association pour l'Etude Taxonomique de la Flore d'Afrique Tropicale (AETFAT)". A number of vegetation maps followed but they did not deal specifically with Africa. The most recent Unesco/AETFAT/UNSO Vegetation Map of Africa (1:5 000 000) is based on the original AETFAT map and contains very detailed descriptions by Prof. F. White [10]. In spite of its complicated name, the map, which appeared in 1983, offers the most extensive and well-founded information on African vegetation available today. It is based on the morphology of vegetation types and on floristic composition rather than on climatic zones. The map's classification of rainforest includes both ever-green forests and semi-deciduous forests including dry semi-deciduous forests, the last of which had been listed by Knapp [11] as marginal high forests (Fig. p. 28). This text makes use of the Unesco/AETFAT/UNSO map when referring to forest types and distribution if not otherwise specified. The term "rainforest" refers to all tropical moist forests including semi-deciduous types, thus coinciding with the common usage of the word.

Area Estimates of Rainforest in Africa

Estimating the area of African rainforests is no less complicated than the classification of their vegetation. In 1976, the United Nations Food and Agriculture Organization (FAO) conducted a study estimating the global area of tropical

moist forests. The resulting figures were based on the AETFAT Vegetation Map published in 1959. Swiss forest engineer Adrian Sommer differentiated in the study between the natural distribution area of tropical moist forests and the actual forest cover then remaining [12]. According to the figures, African rainforests including those in Madagascar, Mauritius and Réunion originally covered 3.62 million square kilometers, a surface about the size of the combined areas of India, Nepal, Bhutan and Bangladesh. This total area could be further subdivided as follows: 2.69 million square kilometers (74%) in Central Africa, 680 000 square kilometers (19%) in West Africa and 250 000 square kilometers (7%) in East Africa including the islands off the east coast. It is peculiar, however, that Angola with 80 000 square kilometers of tropical moist forest was listed under West Africa. Further, the mosaic of rainforest and savannah vegetation along the edges of the distribution area was included in the category of rainforest. This theoretical area of distribution determined by the study was probably valid until the beginning of this century. Sommer calculated the forest cover remaining as follows: 1 490 000 square kilometers or 55% of the original area in Central Africa, 190 000 square kilometers or 28% of the original area in West Africa and in East Africa 70 000 square kilometers, likewise 28% of the area originally found there. Although Sommer noted that his estimates were based in part on uncertain figures, they did confirm the general impression that forest cover in West Africa had greatly decreased. West African rainforests were clearly suffering from destruction on a larger scale than most other rainforests in the world. They showed a decrease of 72% which was far above the global average of 41.6%. Although East African rainforests showed a similar decrease down to 28% of the original distribu-

Swamp forest lines the shore of the Abi Lagoon in the border region between Ghana and the Côte d'Ivoire. Even before World War I, mahogany logs were floated out of the upper Bia River into this lagoon.

tion, the forest there was of a lesser total area and existed in isolated patches making it more vulnerable in general. Central Africa showed a less dramatic decrease but the figures there were known to be less reliable.

The accuracy of Sommers' figures caused some dispute but after 1976 they left little doubt that tropical rainforests were seriously threatened. The study was valuable if only for having pointed out the seriousness of the situation. More reliable estimates soon followed.

Official Forest Assessment

In 1981, the FAO estimated rainforest distribution in cooperation with the United Nations Environment Programme (UNEP). The "Tropical Forest Resources Assessment Project" was based in part on satellite imagery [6]. But instead of differentiating between tropical moist forests and dry forests, the new study assessed open and closed forest formations as well as undisturbed areas and areas subjected to exploitation. So the multitude of figures resulting from the FAO/UNEP study is mainly oriented towards providing information for timber utilization and cannot necessarily be compared with the categories used on geographical vegetation maps. It would have been useful to have at least distinguished between the closed moist forest types and the closed dry forest types. An especially unfortunate aspect of the study is the classification of rainforest as "productive" or "unproductive". This differentiation was made solely on the basis of accessibility for timber exploitation. Inaccessible regions due to hilly and mountainous terrain as well as protected areas, for example, were placed in the latter category. Defining forests of such ecological value as unproductive may have been merely a poor choice of terminology, but it also reveals a tendency within the Forestry Department of the FAO to overlook the diverse and important other values of tropical forests in favor of timber. The project at least confirmed the decrease in forest cover and the rapid rate at which new areas were being opened up, a situation which upon closer look sheds light on the consequences of timber exploitation.

A careful study of the individual country briefs shows that the category of "closed broadleaved forests (NHC)" in Africa is comprised of almost only moist forest types. This is because of the fact that dry forests in the strict sense of the word are quite rare in Africa. Isolated spots of dry forest were recorded and included in the category NHC in Senegal, Mali, Burkina Faso, the Côte d'Ivoire, the Central African Republic, in Southern Sudan and at the upper Zambezi River. These forests were distinguished by the absence of a continuous grass undergrowth. The extensive miombo forest in Angola, Zambia and Mozambique, the mopane forests in the Zambezi Valley and the species rich baobab forest in the coastal areas of East Africa and Angola, on the other hand, were not listed as "closed broad-leaved forests" but as "mixed forest grassland formations (NHC/NHO)". The Unesco/AETFAT/UNSO Vegetation Map defines these

formations as "woodland" [10]. Still other authors use the term "dry forest" [11]. In short, the FAO/UNEP area estimates for closed broad-leaved forests in Africa include almost exclusively various types of tropical moist forests including lowland and montane rainforests, swamp forests, gallery forests and mangroves (Tab. 3). In West Africa, only 5000 square kilometers of dry forest in the northern Côte d'Ivoire were also included in the statistics. In Central Africa, a similarly small area of dry forest in the Central African Republic was recorded as closed broad-leaved forest.

The new areas estimates by FAO/UNEP show that closed broad-leaved forest covered only 2.14 million square kilometers of Africa in the early 1980s. The greater part, 1.73 million square kilometers (80.8%), was located in Central Africa. In 1976, Sommer [12] had apparently underestimated the remaining area of rainforest in Central Africa. This was probably due to an insufficient extrapolation for Zaire even considering the fact that Angolan rainforest was then still listed under West Africa. It is striking to note that the new figures show a drastic decline in West African rainforests. Whereas West African rainforests had earlier accounted for 19% of the total forest cover, they now accounted for only 8.4%. The forests were obviously declining much more rapidly along the Gulf of Guinea than elsewhere in Africa, a fact supported by figures showing considerable areas of fallow land in West Africa. By 1980, 340370 square kilometers had already been listed as fallow. The sad remains of former rainforest between Sierra Leone and Nigeria covered an area the size of Finland and measured just double the area of forest still remaining (Tab. 3). The FAO area estimates are still considered the most reliable today although they are based on data collected over 10 years ago. In view of the present rate of forest destruction they should be considered with care.

Since 1980, the situation has worsened considerably in at least some West African countries: Among the statistics presented by the FAO/ UNEP project are detailed figures on deforestation between 1976 and 1980. On the basis of the known rates of decline forest loss was then projected for 1985 in the various countries (Tab. 4). The resulting extrapolations are themselves probably out-dated today. In West Africa, especially the Côte d'Ivoire and Nigeria showed high losses as can by seen by comparing Tables 3 and 4. The underlying causes of this deforestation will be dealt with later on.

A Belt of Rainforest across Nine Countries

Flying south along the west coast of North Africa, there is no hint over the Tropic of Cancer that rainforest is just ahead. The yellow sand of the Sahara stretches endlessly over the continent to join the surging waves of the sea. Sparse vegetation can be seen along the entire coast of Mauritania all the way to Cape Verde, the most western point of West Africa. It was here that the French founded Dakar in 1862 for strategic purposes. The city's grid of straight streets betrays

Table 3
Distribution of closed broad-leaved forests ("rainforests") in West and Central Africa at the end of 1980 (in km²).
Source: FAO/UNEP [6].

Region West Africa	Forest (Cat. NHCf)	Fallows (Cat. NHCa)
Benin	470	70
Ghana	17180	65000
Guinea	20500	16000
Guinea Bissau	6600	1700
Côte d'Ivoire	44580*	84000
Liberia	20000	55000
Nigeria	59500	77500
Sierra Leone	7400	38600
Togo	3040	2500
Total West Africa	179270	340370

Region Central Africa	Forest (Cat. NHCf)	Fallows (Cat. NHCa)
Cameroon	179200	49000
CAR	35900*	3000
Congo	213400	11000
Equatorial Guinea	12950	11650
Gabon	205000	15000
Zaire	1056500	78000
Angola	29000	48500
Total Central Africa	1731950	216150

Table 4
Estimated distribution of closed broad-leaved forests ("rainforests") in West and Central Africa at the end of 1985 (in km²).
Source: FAO/UNEP [6].

Region West Africa	Forest (Cat. NHCf)	Fallows (Cat. NHCa)
Benin	410	70
Ghana	16080	65900
Guinea	18700	17500
Guinea Bissau	5750	2300
Côte d'Ivoire	30080	97150
Liberia	17700	56700
Nigeria	44500	90000
Sierra Leone	7100	38800
Togo	2940	2400
Total West Africa	143260	370820

Region Central Africa	Forest (Cat. NHCf)	Fallows (Cat. NHCa)
Cameroon	175200	52800
CAR	35600	3300
Congo	212300	12000
Equatorial Guinea	12800	11800
Gabon	204250	15750
Zaire	1047500	85500
Angola	26800	50000
Total Central Africa	1714450	231150

* including 5000 km² of dry forest in the Côte d'Ivoire and an area in the CAR

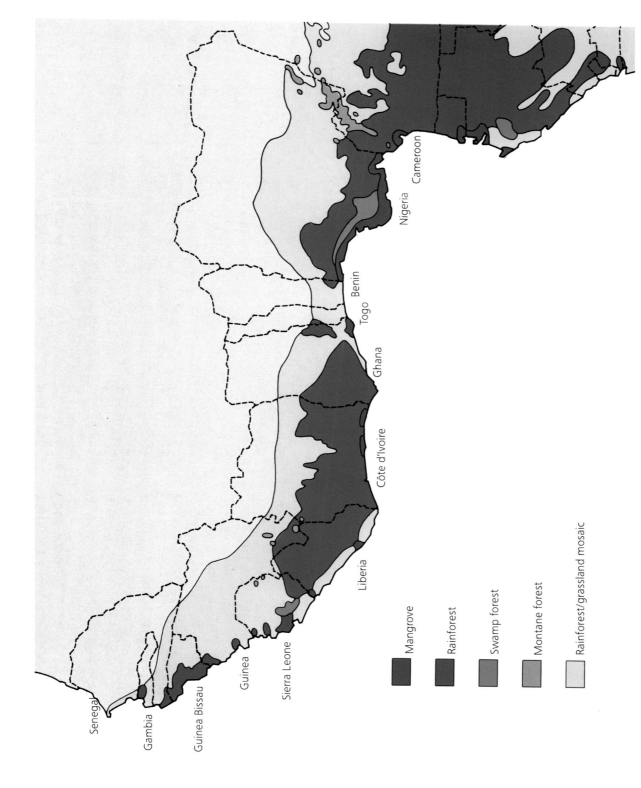

Senegal

Gambia

Guinea Bissau

Guinea

Sierra Leone

Liberia

Côte d'Ivoire

Ghana

Togo

Benin

Nigeria

Cameroon

Mangrove

Rainforest

Swamp forest

Montane forest

Rainforest/grassland mosaic

Rainforest distribution in West Africa, modified according to the Unesco/AETFAT/ UNSO Vegetation Map of Africa [10].

the intentions of it founders: Today's capital of Senegal was to have been the hub of a French-West African dynasty.

The first signs of Africa's moist tropical climate can be seen 200 kilometers to the south at the mouth of the Gambia River where a 640 square kilometer area of mangroves comes into view. Gallery forests along the river's shore and isolated patches of high forest beyond the mangroves mark the transition from savannah to rainforest. The mangroves are especially evident at the mouth of the Casamance in southern Senegal and along the fringed coastline of Guinea Bissau. Guinea Bissau, with its 2430 square kilometers of mangrove forest reaching ten meters in height, provides an important breeding ground for marine fish. Stretches of mangroves can also be found along the coasts of Guinea and Sierra Leone, followed by the beginning of the actual rainforest zone not far from the Liberian border (Fig. p. 35).

A belt of rainforest up to 350 kilometers wide stretches from the eastern border of Sierra Leone along the coast through Liberia and the Côte d'Ivoire all the way to Ghana. In the middle of the Côte d'Ivoire a V-shaped notch cuts into the rainforest from the north. The so-called Baoulé-V ends about 100 kilometers short of the coast. In Ghana, the rainforest zone gradually dissipates near the Volta River. Following a 300 kilometer stretch of savannah – the Dahomey Gap – the rainforest continues in eastern Benin through southern Nigeria reaching at most 200 kilometers inland. The immense delta of the Niger River is overgrown with mangroves covering an area of 9730 square kilometers. West Africa officially ends at the border to Cameroon where the coast runs south but rainforest continues uninterrupted through Cameroon and into the Congo Basin. Realistically, the border to Central Africa could be drawn in Cameroon along the Sanaga River. As we will see later on, the distribution of many West African plant and animal species ends at the Sanaga River. For this reason, occasional reference will be made to rainforests in West Cameroon. The FAO/UNEP study divides West Africa and Central Africa along a national border for practical purposes. All figures for West Africa therefore officially refer to the nine countries from Guinea Bissau to Nigeria. West African rainforests, swamps and mangrove forests cover just one quarter the total area of the countries in question. Only one – Liberia – lies entirely within the rainforest zone. The area of distribution shown in the figure on p. 35 covers about 520000 square kilometers which corresponds to the sum of the areas listed under forest (NHCf) and forest fallow (NHCa) in Table 3.

West African rainforests thus form two major blocks west and east of the Dahomey Gap. The western forest block accounts for 3/4 of West Africa's tropical moist forests. Included in this block is a patch of forest some 10000 square kilometers in area located in the hilly border region between Ghana and Togo. The western block stretching from Guinea Bissau to Ghana is also known as the Upper Guinea forest block.

The term "Guinea" was originally used by early Portuguese travellers to Africa. They in turn had borrowed the word from Moroccan Berbers who called the land south of the Sahara "Akal n-Igui-nawen" which literally means "Land of the Blacks." In geographical terms today, aside from the country name Republic of Guinea, the word generally denotes forested areas on the west coast of Africa.

The Refuge Theory

Many concerned writers have claimed the earth's rainforests had thrived for millions of years until today's consumer let loose the present rampage of destruction. This author does not wish to diminish concern for the rainforests, on the contrary, but it must be recognized that

Distribution of moist tropical forests in Africa over the past 20000 years according to Hamilton [15].

18000 B.C.

6000 B.C.

present

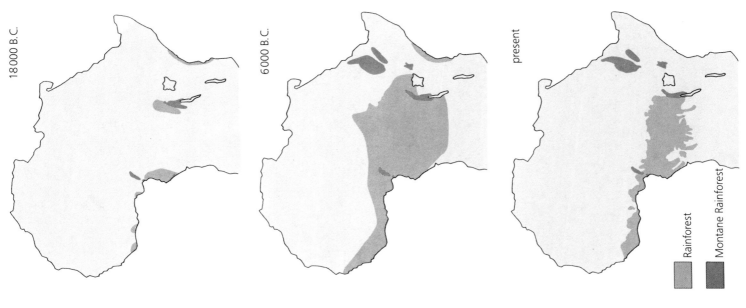

Rainforest

Montane Rainforest

most tropical moist forests are relatively young when seen in the context of the earth's history. According to present knowledge, certain areas of rainforest have existed only for 10000 to 12000 years. This is especially true for most African rainforests. Compared to Southeast Asia or the Amazon Basin, equatorial Africa is a dry continent. And its climate has influenced the distribution of rainforests throughout history. The area of tropical rainforests in Africa has thus varied over time. There is much evidence for the fact that the forests waxed and waned with climatic fluctuations and shrank to relatively small areas during the driest periods. Present-day literature somewhat unfortunately refers to the small patches of forest which existed during the ice ages as "refuges". Such terminology may create the impression that animal species could have actively sought refuge in these remnant forest areas. But each phase of shrinkage or expansion took place over hundreds or thousands of generations allowing species distribution to gradually adjust to the changing forest cover.

Until recently, scientists believed there had been only three to five ice ages in the latter part of the Quarternary, a period reaching back about one million years. It was assumed that the four major glacial periods in the northern hemisphere – Günz, Mindel, Riss and Würm – had come and gone within clearly defined time periods between which interglacial episodes accompanied by a milder climate occurred. This has proven to be much too simple a theory. The examination of deep-sea sediments which have slowly and regularly accumulated over time has shown that there must have been at least nine glacial periods of varying length and severity during the Quarternary. Past climatic fluctuations, however, still remain a subject of debate among the experts.

Scientists generally agree though when considering the past 30000–40000 years. About 18000 years ago, the last severe ice age climaxed in the temperate zones. It had begun some 21000 years ago and slowly came to an end 12000–10000 years before our time. The climate of tropical Africa was cooler and drier during those 10000 years [13,14]. It is generally assumed today that the main effect of the ice ages on the tropics was to make the climate drier and more seasonal. In the 1960s, scientists still believed that extensive periods of rain (pluvial periods) had been the main characteristic of the glacial periods in tropical regions. Today, however, the pluvial theory has been rejected. What was Africa like during the cool, dry periods of the Quarternary? During the height of the last glacial period some 18000 years ago, glaciers even covered the slopes of Africa's mountains. Immense moraines and deeply cut valleys attest to the presence of glaciers on Ruwenzori, Mount Kenya and Kilimandjaro during the earlier Quarternary as well. It was most likely too cold during the ice ages for rainforests to surround these mountains as they do today. It was not until 15000 years ago that glaciers began to retreat from Ruwenzori. Pollen analysis proves the occurrence of tropical forests at Ruwenzori only about 12500 years ago [14]. Even low-lying areas could not have supported large areas of rainforest vegetation before then. The changing climate of the Quarternary must have led to complicated changes in the surface distribution of the rainforests. And there are no indications that the climate was ever stable for longer periods compared for example with the life span of individual trees. The Kalahari sands beneath most of the rainforests of the Congo Basin are impressive proof of the dry glacial periods which occurred in Africa not too long ago. Between 50000 and 10000 years ago, it is very likely that sand dunes blew across land covered by rainforest today [14]. Dry periods with sparse forest vegetation probably characterized the larger part of the Quarternary. During the intermittent warm and moist periods, however, the rainforest expanded, sometimes covering a larger area than its present distribution (Fig. p. 37).

In spite of the drier climate during the last glacial period, forest refuges survived in places where the level of rainfall today measures at least about 2000 millimeters annually. Under drier conditions these areas received just enough precipitation to sustain a moist forest vegetation. The apparent coincidence of maximal precipitation today and forest refuges during the glacial periods leads to the assumption that firstly, wind conditions have not changed drastically since then and, secondly, that the level of precipitation primarily determined forest distribution. Average temperatures were also a few degrees lower than today but it is uncertain as to whether temperature had much influence on forest distribution.

The size of those former refuges is also quite uncertain. Pollen diagrams have given us first hints as to their area. But refuges in West Africa have not been studied particularly well and must still be considered hypothetical. Perhaps the best argument for the refuge theory lies with today's distribution of plants and animal species. Although forest cover has increased over the past 15000 years, many species have not extended their range much beyond the boundaries of the old forest islands. These inhabitants of the forest with their limited distribution areas are living proof of the refuges of long ago.

Centers of Endemism

One would assume that a continuous stretch of lowland rainforest would show just as continuous a distribution of animal species. From Sierra Leone to the Dahomey Gap and again from Nigeria to Rwanda, one might expect a relatively uniform distribution of rainforest fauna. Contrary to this expectation, an outstanding characteristic of rainforests is the surprisingly limited distribution of many species. In some areas even small

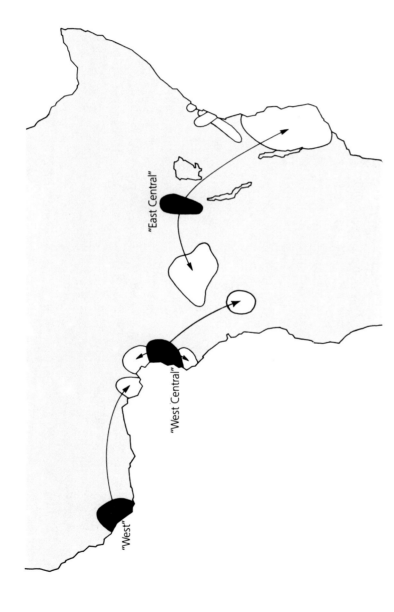

Centers of endemism for rainforest mammals according to Grubb [17]. The arrows connect the centers of highest species diversity (black areas) with secondary centers which probably developed from the major centers.

"West"

"West Central"

"East Central"

streams appear to hinder species from expanding their habitat. An apparently arbitrary line drawn across an otherwise undivided patch of forest seems to keep primates, ungulates and even some species of birds and butterflies from dispersing past the bounds of their range. No where else on the mainland are so many species found with such narrowly defined areas of distribution. These so-called "endemic" species are a unique characteristic of many rainforests throughout the world. In comparison, related species in the savannah have much wider areas of distribution covering various types of savannah habitat. Zoologists were thus the first to raise questions about the cause of endemism and came upon the idea that there might be historical reasons for the distribution patterns in today's rainforests.

Angus Booth was one of the first zoologists to investigate patterns of species distribution in

Africa. He did research at the University College of Ghana in the 1950s and published a number of leading papers on the distribution of mammals in West Africa before his sudden death at the early age of thirty. Although Booth still adhered to the pluvial theory, he correctly assumed that present-day patterns of mammal distribution were related to climatic fluctuations during the Pleistocene period [16]. After his death, Booth's work was continued by Peter Grubb, another British zoologist working in Ghana. Grubb superimposed distribution patterns of mammal species upon a single map. He noted in particular the patterns of endemic species – mammals with limited distribution ranges. In this manner, Grubb discovered that certain areas were especially rich in endemic species and called them "centers of endemism" (Fig. p. 39). These centers show a considerably higher species diversity than surrounding forest areas.

Diversity decreases steadily however, the further one moves from the core of an endemism center. 63% of the forest-dwelling mammal species in Africa occur within the region of a single center of endemism [17]. Interestingly enough, these pockets of species richness are found exclusively in zones of maximal precipitation and correspond with the forest refuge areas of the former cool and dry periods. The colobus monkey quickly moving through the branches of the forest canopy, or the duiker antelope darting through the underbrush of the rainforest both serve to remind us of the drastic climatic events in Africa tens of thousands of years ago! The patterns of distribution for mammal species in the Upper Guinea forest block indicate that the rainforests in the eastern Côte d'Ivoire and in Ghana must have been populated by species coming from the former refuge area. The rivers Sassandra and Bandama appear, however, to have posed insurmountable obstacles for certain mammals, for example the banded duiker.

Even some bird species had difficulties although it is hard to imagine many hindrances to bird migration. Closely related species of passerine birds are often found to have clearly distinct areas of distribution in the centers of endemism west and east of the Central African forest block or even in the Upper Guinea forest block. On the other hand, no passerine species is limited exclusively to the interior of the Congo Basin [18]. Although some song birds have spread far beyond the boundaries of the old refuges, birds in general confirm the same centers of endemism indicated by mammal species.

Some authors divide the Upper Guinean center of endemism into two parts: A western part in Liberia and the western Côte d'Ivoire and an eastern part in the eastern Côte d'Ivoire and western Ghana. It is certainly possible that these two areas were once separated during drier climatic periods when the savannah of the Baoulé-V extended all the way to the coast. Judging from the number of endemic species, however,

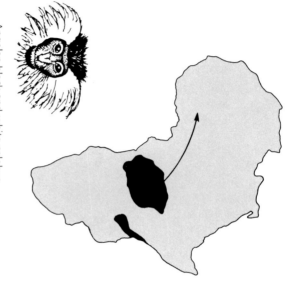

Angolan black-and-white colobus
(*Colobus angolensis*)

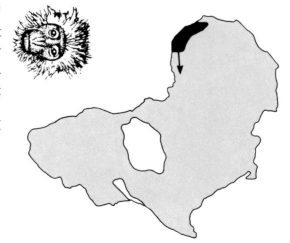

Western black-and-white colobus
(*Colobus polykomos polykomos*)

Guereza
(Colobus guereza)

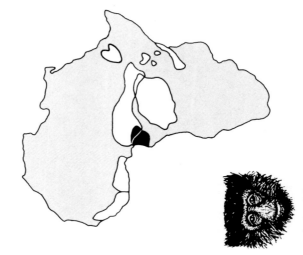

Black colobus
(Colobus satanas)

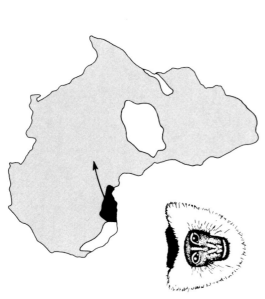

Western black-and-white colobus (ssp.)
(Col. pol. vellerosus)

Distribution of black-and-white colobus species in the order of their probable evolution through isolation in forest refuge areas according to Grubb [17]. See text for details.

there is little proof of a separate center of endemism having existed in the border area between Ghana and the Côte d'Ivoire. There are no mammal species found there which do not also occur in the western part of Upper Guinea. Only some subspecies are endemic to the area. Not even the Dahomey Gap served to deter certain passerine birds from moving out of the Upper Guinea forest block. They are found on both sides of the arid gap all the way to the Niger River and some even up to the Sanaga River in Cameroon. Only there do they give way to a closely related species. Such species distribution patterns indicate that the Dahomey Gap may have been closed until a relatively short time ago.

Actually, botanists should also have wondered at limited patterns of distribution. Most of the some 8000 plant species which occur in West and Central African rainforests show traits of endemism. Nevertheless, most botanists dealing with African flora simply accepted the patterns of distribution and devoted themselves to classifying species and vegetation types. The checklists of plant species for protected areas show a much higher degree of diversity in centers of endemism than in other areas. In the Taï forest area of the western Côte d'Ivoire, for example, 54% of the 1300 plant species identified are endemic [19]. They occur only in this region of Africa. Despite endemism of plant species, they did not all remain static. The forest was obviously capable of expanding considerably beyond refuge areas. Warmer temperatures and higher levels of precipitation during the interglacial periods led to the spread of rainforest far beyond the confines of the refuge centers within only a few thousand years. This relatively fast pace of growth could only have been possible by adaptable tree species with short cycles of generation – the duration between germination and seed production of the adult plant. Such pioneer species are the first to colonize new areas. They

usually have small seeds dispersed by animals [13]. A high percentage of pioneer species may therefore be expected to characterize historically young forests.

For conservation purposes centers of endemism are of very special importance. Large protected areas within these zones harbor the greatest number of plant and animal species. They represent the oldest rainforests existing today and are a proven source for the recolonization of other forest areas with plant and animal species.

The Evolution of Rainforest Species

Although the cool and dry periods during the Quaternary had catastrophic consequences for much of the rainforest ecosystem in Africa, the less favorable conditions also gave rise to new species : The separation of patches of forest into refuges led to the isolation of animal and plant populations. Species either became extinct or they survived and not seldom did they develop into subspecies or even entirely new species. The alternation of cool, dry periods with warmer, moist conditions allowed the newly developed species and subspecies to spread into wider areas. At times they even came into contact with past relatives which had survived in other refuges. We can thus assume that more new species evolved during the climatic fluctuations of the Quaternary period than during the preceding climatically more stable Tertiary [20]. Song birds and smaller species of mammals may generate new species within a relatively short span of time, i.e. in less than 10000 years [21]. Tree frogs (Rhacophoridae) are also known for rapid species evolution. There are 23 species of tree frogs found in the rainforests of West Africa of which only three occur both east and west of the Dahomey Gap. The Baoulé-V, which once divided the Upper Guinea forest block in today's Côte d'Ivoire, similarly led to the creation of distinct species east and west of the area once

Only a small part of Africa's rainforests survived the cool, dry periods during the Pleistocene. Forest refuge areas probably persisted only in the regions of maximal precipitation, regions where annual rainfall today still measures at least 2000 mm.

covered by savannah vegetation. Significantly, the most recent and most flexible form of tree frog, the genus *Hyperolius*, shows the highest degree of species separation by the Baoulé-V and the Dahomey Gap [22].

The evolutionary history of certain species groups of African primates can only be explained by assuming periods of isolation in different forest refuges. The species group of black-and-white colobus monkeys is a typical example of this phenomenon (Fig. p. 40). At one stage in history during a phase of expanded forest cover, the Angolan black-and-white colobus (*Colobus angolensis*) made its way from the center of the continent to West Africa. During a dry period which followed, the Western black-and-white colobus (*Colobus polykomos*) developed in the isolation of the Upper Guinea forest refuge. At a later stage during which a warmer and moister climate had allowed the Dahomey Gap to close again, the newly formed species made its way back east. As the rainforests again retreated to smaller patches, the guereza (*Colobus guereza*) evolved. It has since spread across Africa to Ethiopia and into East Africa having developed a number of subspecies. Somewhere along the way a subspecies of the Western black-and-white colobus developed, *Colobus polykomos vellerosus*, which occurs from eastern Côte d'Ivoire through the patches of forest in the Dahomey Gap almost up to the Niger River. But it is also possible that this subspecies developed during a more recent dry period when the Baoulé-V reached all the way to the coast. The origin of the black colobus is less clear. It may well be that this species already existed earlier in the area of the "West Central" center of endemism.

Aside from the black-and-white colobus, Peter Grubb also investigated the speciation of mangabeys, mona monkeys and red-footed squirrels [17]. Many of these species groups show a similar pattern of distribution: The more they are dependent upon closed rainforest, the more typical their distribution. The surprising similarity among species groups is strong proof for the refuge theory and indicates that species evolution is far from a random process.

∨

A natural clearing in the Krokosua Hills of Ghana, otherwise thickly covered with rainforest. The clearing is a relic of drier climatic periods. Bauxite may be exposed on the surface and the soil is too shallow to support large trees. A species of herb (*Aeschynomene deightonii*) attests here to past ice ages, it otherwise occurs only in Sierra Leone and in Guinea [28]. Forest-dwelling people consider such peculiar forest clearings to be "Juju" places inhabited by the spirits of the forest.

Climate and Soil Determine Forest Type

∨ At certain places in the forest, giant trees practically stand upon bare rock – the ancient basement rock which underlies most of the African continent.

Most people imagine the rainforest to be hot, sticky and swarming with insects. Indeed, many missionaries died of yellow fever during colonial times, a fact which led their fellow Europeans to decry West Africa's shore the "Fever Coast". Today still, a glance at the obligatory innoculations listed on an international vaccination card tends to reassure travellers disembarking from an airconditioned plane in Monrovia or Accra. Tropical disease comes to the minds of many as they find themselves drenched with sweat merely waiting to pass through customs. Surely, no one should forego preventive medication against malaria nor the usual vaccinations but the stories of the "Fever Coast" originated at a time when the rainforest was still considered "green hell" by most non-Africans. Admittedly, it can be uncomfortably hot and humid in West African cities, in areas cleared of forest cover and where plantations have been established.

A Comfortable Climate – in the Forest

Although the relative humidity in closed rainforest is usually around 90% both day and night, temperatures are moderate and quite stable. West Africa's mean annual temperature is 26°–27°C whereas the monthly average ranges from 24°–28°C. The seasonal fluctuation of the mean temperature is therefore considerably less

Variations in air and soil temperatures during a 24-hr period, in closed rainforest and in a clearing. Air temperatures were recorded 1.5 m above ground, soil temperatures were recorded 2 cm below the surface [66].

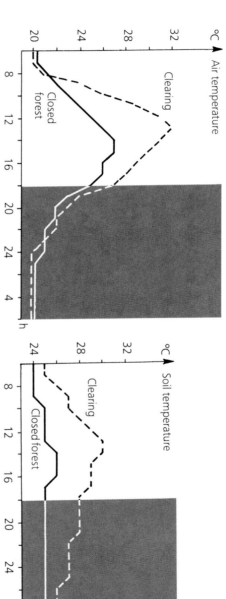

than the day to day fluctuations. Daily temperatures vary greatly, however, depending on where they are recorded: closed forest shows a very different temperature curve than a clearing or plantation area (Fig. p. 48). It can be unpleasantly hot and humid beneath the noon day sun without the shelter of trees above. Microclimatic differences exist not only between closed forest and cleared areas. The same temperature span can be recorded between the floor and the canopy within the forest. Temperatures recorded at canopy level are similar to those measured in clearings at 1.5 meters above ground. Forest vegetation thus greatly influences temperatures near ground level. Consequently, extensive cleared areas show a generally different temperature pattern.

Contrary to popular belief, the climate of the rainforest is quite comfortable for humans. Lowland rainforest is seldom too cold, nor is it ever too hot, so seasonal clothing is unnecessary. It is

quite conceivable that human physiology adapted to a rainforest environment at some past time although archeological findings have not as yet proven this. Temperatures above 35°C in the rainforest are rare and even that is still far below the maximum temperatures reached in subtropical regions or even in temperate climates.

Not only do closed forests in West Africa show steady daily and annual temperatures, their temperature characteristics also show little difference when compared with rainforests at similar altitudes on other continents. This fact strongly indicates the self-regulatory nature of temperature patterns in tropical moist forest ecosystems. The stable temperatures within the forest, however, should not lead to the conclusion that rainforest cannot exist under other conditions. Montane rainforests, in which generally colder temperatures prevail, prove the point. A sufficient level of precipitation is apparently the more

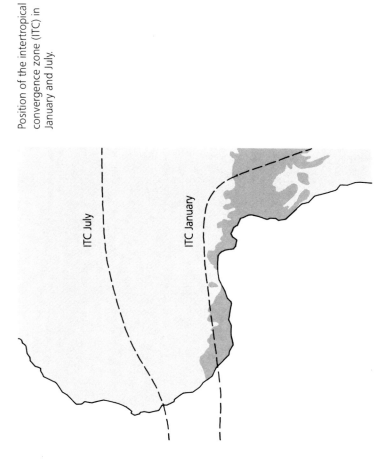

ITC July

ITC January

important factor for the development of rain-forests.

A Lot of Rain – Unevenly Distributed

The forests along the Gulf of Guinea are sandwiched between a maritime and a continental climate. Land air masses meet with oceanic air masses thereby causing more seasonal fluctuations in climate than is the case in other rainforest areas. The line along which these two air masses converge corresponds to the equatorial trough of low pressure encircling the earth near the equator. Over the ocean, it typically runs north of the equator. Scientists have termed this line the intertropical convergence zone (ITC) [23,24].

In the first half of the year, the ITC moves inland with the overhead sun and parallel to the West African coast (Fig. p. 49). At its southern most position, the ITC lies just behind the coast be-

The ITC allows moist, warm oceanic air to move in from the coast and spread across the continent. Precipitation in West Africa is subject to relatively strong seasonal fluctuations.

In July, the intertropical convergence zone lies well over the interior of the continent. But heavy precipitation only occurs much further behind the front, in the rainforest zone.

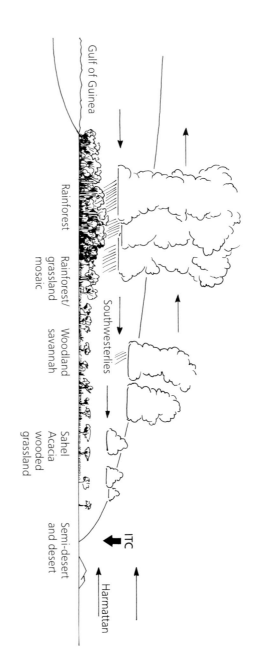

Gulf of Guinea — Rainforest — Rainforest/grassland mosaic — Woodland savannah — Sahel Acacia wooded grassland — Semi-desert and desert — Southwesterlies — ITC — Harmattan

tween 5°N and 7°N. It moves inland at an approximate speed of 160 kilometers a month, reaching its most northern position by July or August at some 20°N far into the Sahara. The northern displacement of the ITC allows warm, moist air to move in from the ocean and spread across West Africa pushing the dry and dusty desert air out of the lower atmosphere. This does not necessarily produce rainfall in all areas under the influence of maritime air masses. Rainfall generally occurs behind the front where maritime air masses are at least 1500 meters thick. This explains why there is hardly any rainfall over the southern parts of the Sahara although moist air reaches the desert. During the second half of the year, the ITC moves back south much more rapidly and dry desert winds blow across West Africa clouding the air with dust and sand particles. Throughout West Africa, desert winds are known as *Harmattan*. They pass through the rainforests of West

Africa around the turn of the year, reaching almost down to the coast and causing a number of smaller forest streams to dry up for some time. The Harmattan cause the relative humidity within the forest to drop from its normal level of 90% to about 70%. At the northern edge of the rainforest, humidity can even sink to desert-like levels [25]. Thus, the climate of West Africa consists of two major seasons: a wet season and a drier season depending on the type of air mass present in the atmosphere (Fig. p. 50). These circumstances explain why coastal regions receive more precipitation than areas further inland. Rain can be expected to fall more frequently where maritime air masses are present for a longer period of time. It is more difficult, however, to explain the varying levels of precipitation from place to place along the coast itself: 3000 mm or more rain fall annually near the coast in Guinea, Sierra Leone and Liberia as well as further east along a narrow stretch of the

Along the small Hana River in the southern part of Taï National Park, Côte d'Ivoire. West Africa's coastal rainforest areas receive the heaviest rainfall, the soils are more severely leached and the trees are not as tall as further inland.

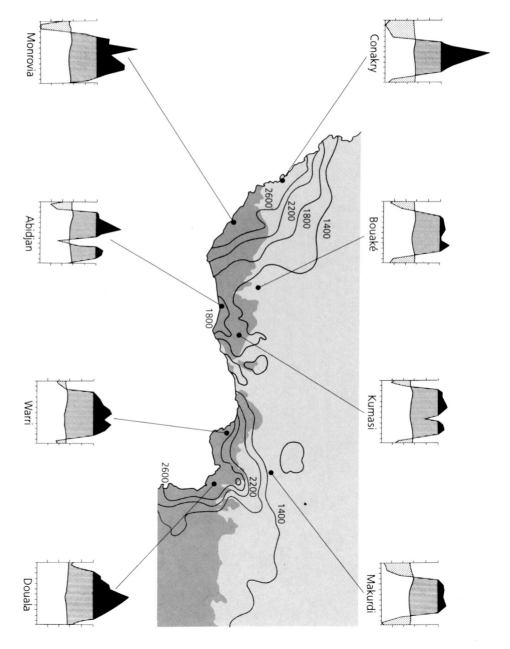

Geographical and seasonal distribution of mean annual precipitation in West Africa's rainforest zone (dark gray areas).

The isolines shown on the map refer to mean annual rainfall in mm.

Climate diagrams: The curve of monthly means of precipitation in the rainforest zone shows two peaks. The black area indicates the perhumid period of the year during which rainfall measures more than 100 mm a month (scale reduced by a factor of ten). The hatched areas indicate the humid period. The lower limit is set by the monthly means of temperature curve which lies between 25° and 30°C in all diagrams. Dotted areas indicate arid seasons.

Niger Delta in Nigeria and in Cameroon. Africa's highest levels of rainfall have been recorded at the foot of Mount Cameroon where some places show impressing levels of 10000 mm and more.

Situated between the two wettest areas along the West African coast is the Dahomey Gap, a dry region where less than 1200 mm of rain fall annually. And even this minimal precipitation occurs rather seasonally which means that the climate is even drier for a good part of the year. Rainforest vegetation cannot survive under such conditions. But how did this abnormality of cli-

mate come to be? Beginning near Axim in western Ghana, the West African coastline runs southeast. Annual rainfall here measures 2032 mm. The coast then turns northeastward at Cape Three Points and in Takoradi – a mere 60 kilometers from Axim – only 1194 mm of rainfall are recorded annually. Takoradi marks the beginning of the Dahomey Gap on the coastline. Since the moist air masses originate from the southwest, they move parallel to the coast from eastern Ghana to Benin producing an effect similar to the rain shadow which often occurs on the lee side of mountains. Although the land behind the coast is flat, little rain falls along this part of West Africa. That is perhaps the most plausible explanation for the arid climate of the Dahomey Gap although there are a number of other theories [24].

Wet and Dry Seasons

Most African rainforests receive an annual precipitation of between 1600 and 2000 mm. As a general rule, levels lower that 1600 mm are typical for transition areas to Guinea savannah vegetation. The quantity of precipitation is an important but not the only determining factor for rainforest growth. Just as important is the seasonal distribution. No less than 100 mm of rain should fall during at least nine months of the year. Thus, if rainforest is to thrive, rainfall should be as evenly distributed as possible over 12 months. West Africa's climate, however, is relatively seasonal. Two extreme examples show what this can mean for rainforest vegetation:
1. Although the coast of the Republic of Guinea receives an annual rainfall of more than 4000 mm in certain places, there is practically no rainfall during the four months from December to March (see Conakry's climate diagram p. 52). The long dry season does not allow rainforest to establish itself here in spite of the high level of annual precipitation.
2. Only 1230 mm of rain fall in Ibadan (Nigeria) per year and yet the city lies within the rainforest zone. The dry season is shorter than three months and the high relative humidity remains constant throughout the year.
It is peculiar to note that the West African rainforest zone shows two distinct periods of high rainfall separated by a short dry season (Fig. p. 52). Surprisingly, during July and August when the ITC is at its furthest inland position and the moist air masses over the coast are at their largest, very little rain actually falls. This drop in rainfall lasts about six weeks and is known as the

"little dry season". The reasons for this phenomenon are not easily explained. Climatologists have found that in July and August the ocean winds change direction from southwest to west and thus blow more or less parallel to the coast. So the winds approach the rainforests east of Sierra Leone from the continent and not from the sea thus producing a rain shadow effect similar to that in the Dahomey Gap. West Africa's coastal geography could therefore be responsible for the twofold wet season. But climatologists do not hesitate to cite other theories for this unusual phenomenon [24]. For example, behind a maritime front cooler air is often trapped near the ground under warmer air above. Such inversions tend to calm weather conditions.

In addition to the level of precipitation and its seasonal distribution, the relative humidity of the air and the ability of the soil to retain moisture are also probable factors in determining rainforest cover. There is still too little known about their precise effects, but they, in turn, are also dependent upon rainfall.

Yellow and Red Soil upon Ancient Rock

Northern and western Africa consists of a low-lying plain of immense size reaching far into the center of the continent. Except for the Atlas, the Hoggar and Tibesti massifs, few summits rise more than 1000 meters above sea level. Only the Nimba mountains rising up to 1700 meters stand out in an otherwise flat or undulating landscape where the borders of Liberia, the Côte d'Ivoire and the Republic of Guinea meet. In the east, the West African plateau jutts against the volcanic highlands of western Cameroon which end with Mount Cameroon directly on the coast and surface again to form the islands Fernando Po, Principe and Sao Tomé.
Although the land rises gradually from the low-lands along the Gulf of Guinea, only in very few

This former logging road in southern Ghana has become a major route of transportation. The ferralitic soil quickly erodes, however, and makes expensive maintenance work necessary.

places does it reach 300 meters above sea level. West African rainforests are thus almost exclusively lowland rainforests with only slightly hilly areas.

The ancient crystalline shield underlying most of the African continent is exposed at various places in the rainforest. Granites, gneisses, quartz and schists are the basis from which rainforest soils develop by intensive weathering processes. Although the soil is deep, nutrients are stored only in the thin topsoil layer. In some places, outcrops of granite protrude from beneath the surface creating peculiar rock formations at times reaching far above the surrounding forest canopy.

In the wettest areas along the coast, weathering of the basement rock has produced ferralitic soils of a yellowish hue. They are quite acidic and leached out especially in areas of maximum precipitation. Few minerals remain. Aside from kaolinitic minerals with iron and aluminium oxides, the soil is extremely poor in nutrients. Rainforest soils of this kind, also known as oxysols, are particularly unsuitable for agriculture. Further inland where precipitation measures about 1500–1750 mm annually, ferralitic soils

Moist semi-deciduous rain-forest, about 160 km from the coast. Some of the emergent tree crowns have shed their leaves during the dry season.

are of a reddish color and less acidic or even neutral. These are the forest ochrosols. They are richer in nutrients especially in less washed out places, in depressions and at the bottom of valleys. In transitional forest areas bordering on savannah and in savannah regions, ochrosols are of a reddish brown color. Soils of this type are also known as savannah-ochrosols. The term "laterite" and to some extent also the term "lato-sol", words earlier used to describe tropical soils in a general sense, have led to considerable confusion and are now often replaced by the above terminology.

Not only does the underlying rock influence the soil type formed above, but the amount of precipitation obviously also plays a role by determining the soil's nutrient content. Rainfall thus has a twofold effect on forest vegetation: Directly, by determining the amount of water available to the plants, and indirectly, by leaching the soil of its nutrients and so influencing its fertility. Relatively fertile forest soils are found on late tertiary volcanic rock formations. The soils which have developed from these basalts have a fine, claylike texture and are well-known among farmers as an ideal basis for agriculture. This type

Attempts at clearing and cultivating here will not go unpunished: the buttresses of this ancient tree stretch over bare granite!

of soil is common in the highlands of western Cameroon but otherwise quite rare in West African rainforests.

In general, the rainforest has never been more wrongly judged than concerning its soil. The lush and immensely diverse vegetation of tropical rainforests was long thought to be partly due to fertile soils. It seemed difficult to believe that tropical rainforests could exist on surfaces almost bare of nutrients. Consequently, many a well-meaning development project based on the false assumption of fertile soil has led to failed harvests, fallow land and erosion. Only

recently has increased public concern for the fate of the rainforests led to more careful planning of development projects and a better understanding of soil conditions.

The Result:
Different Types of Rainforest

The complex interaction of various environmental factors determine the distribution and characteristics of the rainforest. Climate, geology and soil are the most important although no single factor shows a perfect correlation with the distribution area of one specific type of forest. The level of precipitation, however, is of central importance in determining the various forest types [26]. Since rainfall decreases as we move inland from the very wet coastal areas, the structure of the forest changes accordingly. It is commonly believed that the tallest trees grow in those rainforests where the most rain falls. But that is neither the case in the Amazon region nor in Africa. Both the biomass as well as the girth and height of individual trees are less in the wettest tropical rainforests than in evergreen and semi-deciduous forests with more seasonal climates. This could be due to leached soils and an accordingly low level of nutrients. Increased cloud cover leading to a generally lower light intensity could also be responsible for the lower growth potential [26]. Thus, the giant trees for which the rainforest is famous do not characterize the wettest evergreen forests but rather a drier, more seasonal forest type.

Unfortunately, the Unesco/AETFAT/UNSO Vegetation Map of Africa does not distinguish between rainforest types. The map shows only the approximate distribution of "wetter" and "drier" types and an undefined mosaic of both. Although quite valuable otherwise, this factor has led to the map being declared useless by some. While it is a genuine vegetation map for savannah and dry woody areas, it does not merit the term concerning vegetation within the distribution area of rainforest. It is no wonder that the task was difficult. The Gulf of Guinea is lined with a number of relatively small countries, a fact which has not encouraged a uniform classification of local vegetation and climatic conditions. Not only is communication hampered by poor road connections but the language barriers, too, pose quite a problem. From the Republic of Guinea to Nigeria, the national language alternates between English and French from country to country on an almost regular basis.

Rainforest Types in Ghana

The best analysis of West African rainforest types is undoubtedly that conducted by John B. Hall and Michael D. Swaine in Ghana. These two British scientists, Hall in particular, spent many years at the University of Ghana in Legon, near Accra, carrying out plant sociological studies. Their classification is based on 155 sample plots (25 × 25 meters) distributed throughout Ghana's rainforest. Among other information, they listed all vascular plant species occurring in the plots (i.e. omitting algae, fungi and mosses). Tree seedlings – a common form of vegetation in the undergrowth of rainforests – were also recorded. In order to achieve as complete a picture as possible of the forest in its original state despite the existing degree of forest destruction, most of the plots were chosen within forest reserves and "Juju" places, sacred patches of forest reserved for religious ceremonies.

Of the 1248 vascular plants observed, only the 749 species occurring in at least three different plots were used for classification. For obvious reasons less common species are not as helpful in determining similarities between plots. A computer data analysis technique called "ordination" resulted in a classification of sample areas on the basis of similar characteristics. The groups so classified have ultimately been defined as forest types [27]. Environmental factors such as precipitation and geological criteria were taken into consideration in the analysis.

Hall and Swaine drew up a vegetation map of the rainforest zone in Ghana. It shows four main types of forest according to decreasing levels of precipitation from the coast moving inland (Fig. p. 59). Since Ghana's rainforests are interrupted by the savannah of the Dahomey Gap, the forests towards the east also become increasingly drier. In this area directly on the coast, Hall and Swaine recognized a special "southern marginal" forest type which only exists in small

Major rainforest types in the Côte d'Ivoire. Dotted areas were already more or less damaged by agriculture in the 1970s [31].

Evergreen forest types

Semi-deciduous forest types (see Tab. 5)

Liberia

Man

Abidjan

Dimbokro

Ghana

patches today. In addition, they defined an "upland evergreen" type and the structure of certain isolated forest patches. Hall and Swaine's description of forest vegetation in Ghana explains all the forest types in detail [28]. Although types of forest were defined according to floristic analysis, the authors used climatic and physiognomical terms in their definitions. For our purposes, a presentation of the four main types comparable with those occurring in other West African countries will suffice. The figure on pp. 60/61 shows a schematic cross section of Ghana's rainforest zone at a right angle to the coast near Axim and moving north.

Comparable Types of Rainforest Elsewhere

Hall and Swaine may indeed have used a complex methodology to classify the rainforests in Ghana, but the complexity of the rainforest demands an accordingly detailed analysis. The commonly applied method of choosing more or less arbitrarily one or two characteristic species to classify forests is not adequate when 100 or more species of trees occur in a relatively small area. Clearly, dominant species according to which a type classification can be made are rare in rainforests. One exception in West Africa are the upland forests of the Nimba massif and the

Rainforest types in Ghana according to Hall and Swaine [27,28]. See Table 5 for comparison with other classification systems.

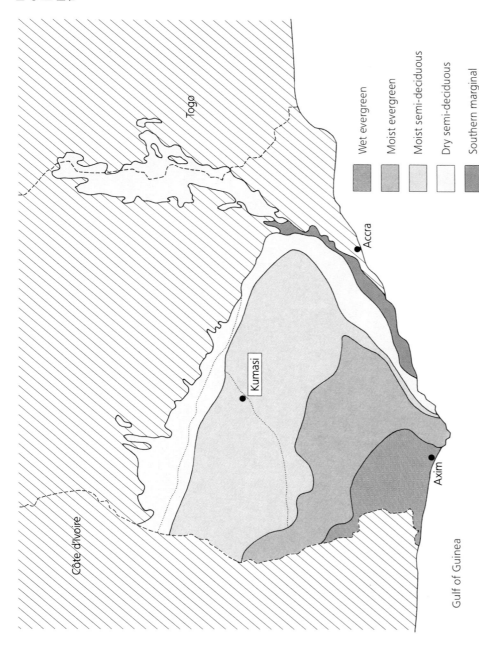

- Wet evergreen
- Moist evergreen
- Moist semi-deciduous
- Dry semi-deciduous
- Southern marginal

Foula Djalon plateau in the western Côte d'Ivoire, where the Guinea plum tree *(Parinari excelsa)* occurs in high densities. Some experts today refuse to accept a classification of rainforests according to purely floristic principles. Most needs are better served by a classification according to a variety of factors such as forest structure, degree of humidity and soil conditions.

Before Hall and Swaine's findings were published in 1976 [27], other more standard methods of plant sociology had been used to classify forest types in West Africa. Attempts to identify types according to characteristic species had al-

ready been made in Ghana early this century. Taylor's classification became especially well-known [30]. Taylor used species of commercial timber to indicate forest type. At the time his work was published in 1952, timber exploitation was the main factor of interest regarding tropical forests, a fact which explains his selection of criteria. Despite the elementary nature of Taylor's plant sociology, a comparison with Hall and Swaine's precisely defined forest types shows a close correspondence with the main types (Tab. 5).

A very detailed vegetation map for the Côte d'Ivoire was published in 1971 [31]. This marve-

Schematic cross section through the rainforest zone in Ghana beginning at the coast near Axim and moving north. Rainforest types from left to right: wet evergreen, moist evergreen, moist semi-deciduous, dry semi-deciduous. See Fig. p.59.

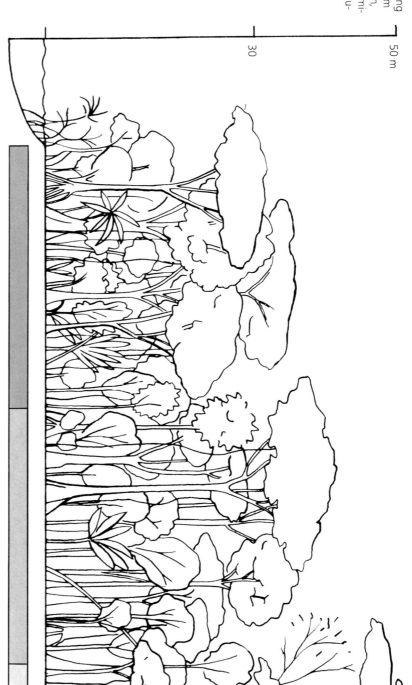

Wet evergreen

Moist evergreen

Wet evergreen type

(cf. Fig. p. 64)

Annual precipitation for this type of forest measures at least 1750 mm and more than 2000 mm at certain locations; it can be considered "rainforest" in the strictest sense of the word. The soils are leached by the heavy rainfall and quite acidic. Plant diversity is at its maximum. Up to 200 species of vascular plants can be found on a 25 × 25 meter sample plot. About 20 are considered characteristic species – they occur almost exclusively in this type of forest. Marketable timber species, however, are not particularly abundant in wet evergreen forests. Only nian-gon, lovoa and azobe are relatively common. The forest canopy is strikingly low and seldom rises above 40 meters. The trunk diameters and crowns of individual trees, however, are similar to those in moist semi-deciduous forests where trees reach greater heights.

Moist evergreen type

Weather stations in forests of this type usually record 1500–1750 mm of annual rainfall. Species diversity is lower than in wet evergreen forests: A 25 × 25 meter plot harbors at most 170 plant species. The number of characteristic species is also fewer. Species of commercially important timber, however, are much more common: African mahogany, sapele, gedu nohor and makore. Obeche, a commercial species which does not occur in wet evergreen forests, appears here. Some trees shed their leaves for short periods of time, but although this type of forest is similar in appearance to moist semi-deciduous forest, the trees are not of the same height. The tallest speci-mens barely reach more than 43 meters.

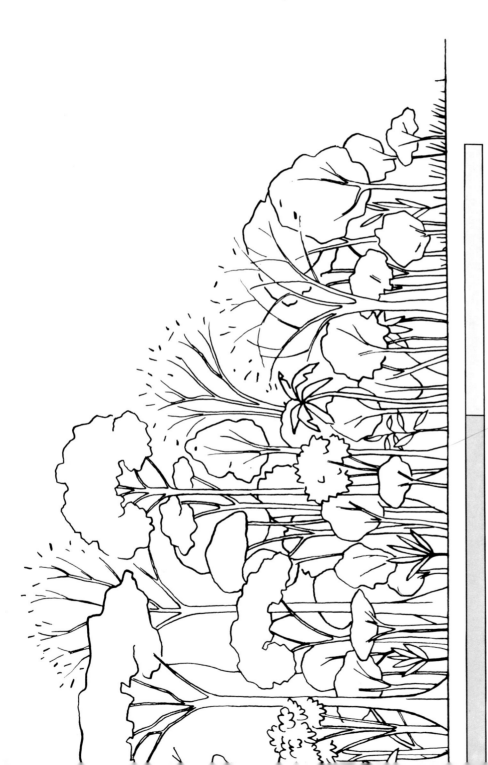

Moist semi-deciduous

Moist semi-deciduous type

(cf. Fig. p. 65)
This type of forest receives an annual rainfall of between 1250 and 1750 mm. Hall and Swaine subdivided this type into a moister, southeastern subtype and a somewhat drier, northwestern subtype. Of all the forest types, this is the most common throughout Ghana and probably covers the largest area in all of West Africa. Species diversity is measured at more than 100 species of vascular plants in a 25 × 25 meter sample plot. This type of forest is particularly rich in commercial timber species and is consequently the most exploited. Timber companies especially favor the reddish woods African mahogany, utile, sapele and makore. Most tropical crops also grow well in this climate. While the upper canopy is dominated by evergreen and deciduous tree species in fairly equal distribution, the lower canopies consist mostly of evergreen species. This type of forest boasts the tallest trees. Certain tree giants may reach up to 50 or 60 meters in height thus earning the forest its common name: high forest.

Dry semi-deciduous

Dry semi-deciduous type

Most of this type of forest receives between 1250 and 1500 mm of rain per year. Northern areas may only receive 1000 mm which is extremely little for rainforest vegetation. Hall and Swaine listed an additional subtype – the "fire zone subtype" occurring along the line of transition from closed forest to savannah in the north. Fires following especially long dry periods are not uncommon here and leave their mark on the forest floor. Larger trees, however, usually remain unharmed. It is here that we find the first two species of trees also belonging to savannah vegetation: *Afzelia africana* and the ebony wood species *Diospyros mespiliformis*. Species diversity is relatively low: A 25 × 25 meter sample plot harbors only 40 – 100 plant species. Commercial timber species are also more seldom, only obeche and iroko are common. The canopy reaches a height of 30 to 45 meters and is no longer fully closed.

The scientific and further commercial trade names for the timber species mentioned above are listed in Annex 1.

Table 5
Comparison of different rainforest classifications in Ghana and the Côte d'Ivoire

Rainforest types in Ghana according to Taylor (1952)	Rainforest types in Ghana according to Hall and Swaine (1981) with annual precipitation	Rainforest types in the Côte d'Ivoire according to Guillaumet and Adjanohoun (1971)
Cynometra-Lophira-Tarrietia "Rainforest"	Wet Evergreen (>1750 mm)	– Fôret sempervirente à *Diospyros* spp. et *Mapania* spp. – Fôret sempervirente à *Eremospatha macrocarpa* et *Diospyros mannii* (partly)
Lophira-Triplochiton Association	Moist Evergreen (1500–1700 mm)	– Fôret sempervirente à *Eremospatha macrocarpa* et *Diospyros mannii* (partly) – Fôret sempervirente à *Turraeanthus africanus* et *Heisteria parviflora* – Variante à *Nesogordonia papaverifera* et *Khaya ivorensis*
Celtis-Triplochiton Association	Moist Semi-deciduous (1250–1750 mm)	– Fôret semi-décidue à *Celtis* spp. et *Triplochiton scleroxylon* (partly)
Antiaris-Chlorophora Association	Dry Semi-deciduous (1000–1500 mm) Fire Zone Subtype	– Fôret semi-décidue a *Celtis* spp. et *Triplochiton scleroxylon* (partly) – Fôret semi-décidue à *Aubrevillea kerstingii* et *Khaya grandifolia*

lously colorful map probably has no equal in all of Africa. Although the map also uses character-istic species to classify forest types within the rainforest zone, it has an important advantage: Not only does the map depict the distribution areas of forest types, it also shows the extent of actually forested areas – areas which were as yet undisturbed in 1971. The majority of the Côte d'Ivoire's rainforests had already been sacrificed to both plantations and small-scale agriculture. Forest destruction has since continued and the lovely vegetation map may soon be nothing more than an historical document reminding us of what the Côte d'Ivoire once was (Fig. p. 58).

Rainforests in other West African countries have not nearly been as carefully studied and classified as in Ghana or the Côte d'Ivoire. Liberia and Sierra Leone are but two examples. Not only have computer-aided methods of vegetation analysis just recently been developed, but many West African rainforests were destroyed before they could be classified. At the Forestry Department of the University of Ibadan a classification of Nigeria's forests was made in the 1970s on the basis of data which had been collected around 1930. More recent data were not available because Nigeria has since lost most of its rainforests. A method of data analysis similar to the "ordination" technique used by Hall and Swaine was also applied. Strangely enough, the Nigerian study was also conducted by a man named John B. Hall [32]. Ibadan's John B. Hall, however, was not the John B. Hall who had worked at the University of Legon in Ghana and died in 1984. In Nigeria, too, forest types in the southern part of the rainforest zone proved to be characteristically moister than in the north. Soil quality was also shown to have an influence on forest type. Unfortunately, the data collected

John B.Hall, one of the most knowledgeable researchers of West African flora, could correctly identify practically every rainforest plant. Until 1980, he was responsible for the Herbarium at the University of Ghana, Legon. In the foreground: a herbacious palm species, *Sclerosperma mannii*.

before World War II were not done so in a sufficiently consistent manner to allow for a precise classification of Nigeria's rainforests.

A consistent study of the rainforests along the Gulf of Guinea, which would allow the use of a uniform terminology for forest types, has never been made. It is thus difficult to compare the forests from country to country. Identical criteria were not even applied for the classification of the Upper Guinea forest block between Sierra Leone and Ghana. And in Liberia, plant sociological studies have never even been conducted. Surprisingly, however, the first botanical collections in Ghana were made in 1697, in Sierra Leone in 1772, in Liberia in 1841 and, finally, in the Côte d'Ivoire in 1882. Despite this early interest, the specimens collected were never used to classify the rainforest but shipped to Europe to satisfy taxonomic interests [33]. Until recently, botanists have apparently not been interested in the forest as a plant community. But aside from that, French-speaking scientists did not, or could not communicate with their English-speaking colleagues and vice versa. Today, it is no longer likely that a sound classification of West African rainforests can be made on a uniform basis. Too much forest has been disturbed or destroyed over the past decades.

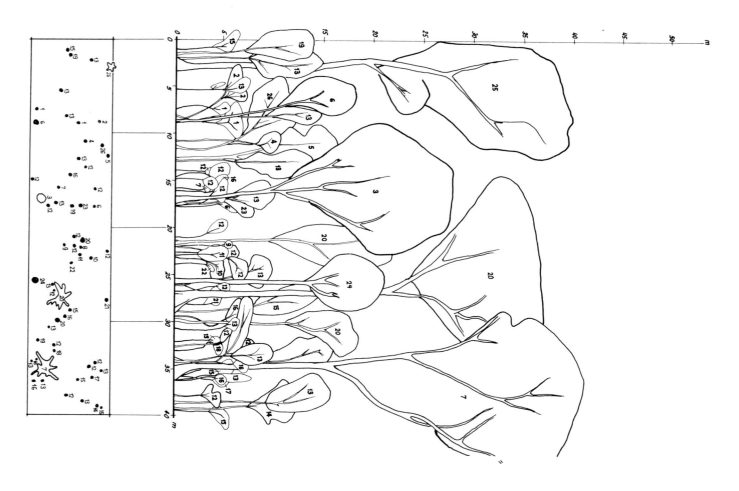

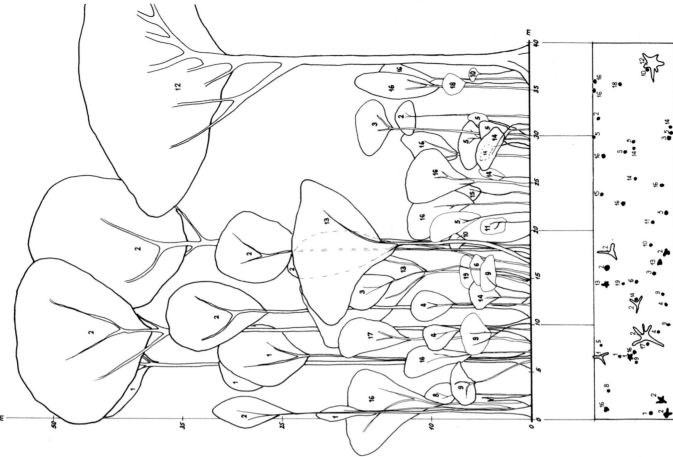

◁

Forest profile of an area
8 × 40 m, recorded in the
wet evergreen rainforest type
in the Ankasa Wildlife Reserve, southwestern Ghana.
Trees and shrubs of less than
5 m in height are not shown.

1 Drypetes leonensis
2 Ouratea reticulata
3 Dacryodes klaineana
4 Chytranthus carneus
5 Scytopetalum thieghemii
6 Trichoscypha arborea
7 Cynometra ananta
8 Tabernaemontana crassa
9 Massularia acuminata
10 Ouratea calophylla
11 Tapura ivorensis
12 Pleiocarpa mutica
13 Drypetes aylmeri
14 Trichoscypha sp.
15 Memecylon lateriflorum
16 Pancovia turbinata
17 Newtonia duparquetiana
18 Placodiscus oblongifolius
19 Aidia genipaeflora
20 Tarrietia utilis (Niangon)
21 Manilkara obovata
22 Cola caricifolia
23 Uapaca corbisieri
24 Diospyros sanza-minika
25 Lovoa trichilioides (Lovoa)
26 Diospyros kamerunensis

▷

Forest profile of an area
8 × 40 m, recorded in the
moist semi-deciduous rainforest type in Bia National
Park, western Ghana. The occurrence of Pterygota species
is amazingly common. Trees
and shrubs of less than 5 m
in height are not shown.

1 Pterygota sp.
2 Pterygota macrocarpa
(Pterygota)
3 Pterygota bequaertii
4 Strombosia glaucescens
5 Baphia nitida
6 Baphia pubescens
7 Drypetes chevalieri
8 Rinorea oblongifolia
9 Isolona campanulata
10 Diospyros mannii
11 Drypetes gilgiana
12 Entandrophragma utile
(Utile)
13 Bussea occidentalis
14 Hunteria eburnea
15 Calpocalyx brevibractea-
tus
16 Corynanthe pachyceras
17 Ochthocosmus africanus
18 Celtis mildbraedi
19 Millettia rhodantha

A Diversity of Plant Life on Poor Soil

P.W. Richards called Africa the "odd man out". The famous botanist, who was the first to describe tropical rainforest in detail, was referring to the plant diversity of African rainforests [34]. Other botanists have also written of the relatively limited number of species on the continent. For all of tropical West Africa, about 7000 plant species have been listed [33] and about 8000 have been counted in the entire rainforest area of both West and Central Africa [10]. Compared to the 8500 plant species supposed to occur in continental Malaysia, a much smaller rainforest area, plant diversity in Africa does indeed appear to be rather limited. Whereas 403 species of orchids occur in West Africa, there are 927 species on the Malaysian peninsula [35]. In Southeast Asia, 100 and often considerably more tree species are found per hectare. There are usually less than 100 in Africa. A study of 23 hectares in rainforests of the highest diversity in peninsular Malaysia revealed 357 different tree species with diameters of 30 centimeters or more. 140 of those species were represented by a single tree specimen [36]. Thus, not only do Malaysian rainforests show a high degree of diversity on the whole, but also a high number of different species are present in small areas. Within 300 square kilometers of Ghanaian rainforest in the transitional area between

the moist evergreen and the moist semi-deciduous zone, on the other hand, only 191 larger tree species are found (Tab. 6). That number may be impressive in comparison with forests in more temperate zones, but some rainforests in Southeast Asia harbor double as many species.

Estimates of species diversity, however, should be considered with care: A rainforest area of comparable size in the southern parts of Korup National Park in western Cameroon was found to contain about 400 tree species [38]. Species diversity can thus vary greatly from place to place and is not a fixed value to be applied to a country as a whole, let alone to an entire continent. Besides, any reliable estimate of species diversity is the result of quite a time-consuming analysis. At least four hectares need to be investigated due to the extremely localized occurrence of many species [26].

Are African Rainforests Inferior?

Why is there less diversity of vegetation in African rainforests than elsewhere? Were too many species lost during the cool and dry periods of the Pleistocene, when glaciers covered much of the northern hemisphere? Or are there other reasons? We can only speculate on the possibili-

∨ Only about 1 % of the light entering the tree crowns above reaches the forest floor. Tree seedlings and the few ground herbs have to make do with twilight and sun flecks.

∨
Among the many African pygmy trees is *Drypetes ivorensis*, a cauliflorous tree which scarcely reaches 5 m in height.

≫
Rattan palms also occur in African rainforests; *Ancistrophyllum secundiflorum*, a climbing palm, reaches some 30 m in height.

ties. Similar cool, dry periods occurred on other continents and also resulted in rainforests retreating to forest refuge areas. Why should the shrinking process have had more drastic consequences in Africa than elsewhere? Other factors may have also played a role: West and Central African forests are almost exclusively lowland rainforests on flat undulating terrain, a circumstance which may have led to a more uniform vegetation. Extreme tectonic structures – high

Table 6
Number of vascular plant species found in the Bia Reserves of Ghana, according to Hall [37]

Plant type	Closed canopy rainforest	Secondary forest	Swamp and riparian forest	Total
Trees taller than 8 m	174	10	7	191
Shrubs and small trees (<8 m)	86	13	1	100
Slender understory climbers	52	36	4	92
Large lianas, reaching the canopy	95	3	4	102
Epiphytes or hemi-epiphytes	43	1	–	44
Parasites	2	–	–	2
Free-living ground herbs	52	46	11	109
Total species	504	109	27	640

mountains and deep valleys, as they occur for example in most Southeast Asian rainforests, encourage species evolution. Plants and animals adapt to different ranges of altitude over the course of their evolutionary history. If populations should become isolated through mountain ranges or valleys, they will continue to develop and eventually become separate species. At some later time, the newly evolved species may again disperse into more extensive areas.

Richards further attributed the limited diversity of African rainforest vegetation to the evolution of man and his use of fire: "Even in the depth of the so-called primary forest, there is often evidence of former human occupation in the form of pottery and charcoal fragments in the soil" [34]. It would be a strange conclusion, however, to thus assume Africa's rainforests consist exclusively of secondary forest disturbed and damaged by man. Firstly, the forest population of prehistoric man was very low, and secondly,

their influence on today's forest vegetation could hardly have mattered since the drastic climatic events during the Pleistocene occurred much later. Man's use of fire in prehistoric times probably dates back to several hundred thousand or even up to one million years ago. The climatic events which last caused the rainforests to retreat to small forest refuges occurred only 18000–20000 years before our time (Fig. p. 37).

Indeed, bits of charcoal are occasionally found buried in the soil of closed African rainforests and indicate more recent human activity. Africans have traditionally hunted forest game in remote and undisturbed rainforest areas. Traces of both ancient and more recent hunting camps can be found along the paths between forest villages as well as in more isolated areas. Today, as in the past, game meat is smoked at the camps to preserve it and make it easier to transport. And as in rainforests all over the world,

The umbrella tree (*Musanga cecropioides*) is the most important example of secondary vegetation throughout African rainforests, here along an old logging route.

forest people have practiced traditional shifting cultivation. The clearings, however, were regularly allowed to regenerate from the fringes of the surrounding forest. But this is no more reason to declassify Africa's rainforests to the status of secondary forest than those in Amazonia or Southeast Asia, where man has lived for hundreds of generations. It would certainly be helpful to clearly define what is to be understood under the term "secondary forest" in order to avoid such confusion. Secondary vegetation is after all a natural phenomenon which occurs wherever a large tree has fallen. Over a few cen-

turies, just about every bit of rainforest is subjected to changes of this nature. After a hundred years or more, the secondary vegetation which has sprung up in clearings or forest gaps reaches a climax and can scarcely be distinguished from the surrounding primary forest [35].

Diversity and Its Surprises

Richards' comments on the limited flora of African rainforests were generally repeated without much question by authors who either had no

With its fan-shaped leaves, the umbrella tree soon casts shadows over forest gaps, without which actual rain-forest trees cannot germinate.

personal knowledge of rainforests in Africa or who compared very elementary studies in Africa with much more detailed investigations in Southeast Asia or Latin America. An exact study of the diversity of plant species in African centers of endemism could lead to some surprises! But whatever the case may be, there are additional indicators of habitat diversity aside from the number of plant species: structural complexity for example, or diversity of fauna. At least in regard to mammal species, African rainforests are not less diverse than rainforests across the globe – on the contrary. But this

chapter does not aim to plead for the biological diversity of African forests. Rather, it warns against preconceptions that Africa is an inferior, botanically devastated continent. The fact is that many African rainforest areas, especially in Central Africa, have never been systematically studied regarding species diversity, neither plant nor animal. New plant and animal species are continuously being discovered. Even new species of higher plants and animals were recently discovered in detailed studies of forests in western Cameroon: An otherwise unknown shrub species was found on the shores of the Ndian River

Fallen leaves and branches decay in usually less than a year. Large trunks take much longer, depending on their position and degree of lignification. One may sink knee-deep in rotting wood.

and given the name *Deinbollia angustifolia (Thomas)*[38]. In 1988, no less than ten new species of fish were discovered in the same river system [39]. In the same year, a British zoologist also stumbled upon a monkey species unknown to science in the rainforests of Gabon. The monkey belongs to the guenons and is distinguished from related species by its strikingly light colored tail. It appears as though bleached by the sun and hence the species name: *Cercopithecus solatus*. The unusual discovery was made in the Forêt des Abeilles, which was until recently an undisturbed area of forest in Gabon. Today, however, timber exploiters have moved into the area [40]. As this example of novelty among primates shows, African rainforests have not yet been sufficiently researched. Present knowledge of forest fauna and flora does not even allow an adequate idea of the diversity of species and their distribution.

Recycling Nutrients

Following a meandering stream through the underbrush of the rainforest, one may be surprised here and there to see sparkling water flowing over bare granite. The water has not eroded its way through deep gullies, no – it flows aimlessly across the forest floor, swishing between protruding roots and disappearing beneath large-leaved plants to finally converge with a stream running over crystalline gravel and light-colored quartz sand. The rivers and streams in the rainforest tell us of ancient Africa, of the crystalline shield of Precambrian age, 600 million years old, underlying most of the African continent. And again, it is the waterways that show us how shallow the soil can be. Sooner or later one begins to wonder where the nutrients come from to nourish such luxuriant vegetation in its strive for light. Obviously, the structure and composition of the forest can not depend only upon the soil. But what mechanism permits a plant in the rainforest to so free itself from the soil? A complete and rapid decay of organic matter on the one hand, together with an efficient up-take of the nutrients thus made available, allow the living plants to thrive in spite of the shallow soil.

Rapid Decay

An array of small organisms infiltrate the decomposing material which basically consists of plant matter. Dead trees, branches, twigs, leaves, flowers and fruits accumulate in large

> The decaying stilt roots of a majestic sugar plum tree (*Uapaca guineensis*).

<< Besides bacteria, termites, wood-lice, millipedes and beetles, fungi play an important role in the process of decay.

quantities in the forest. The proportion of plant matter digested by herbivores and returned to the ecosystem in the form of cadavers is comparatively small. In contrast to temperate zones, earthworms play a smaller role in the process of organic decomposition. There are not many earthworms in the rainforest but those that exist can reach huge dimensions. Giant specimens belonging to the tropical family *Megascolecidae* are rare, however, and it takes a bit of luck to find one. Bacteria, fungi and arthropods are the most important organisms in the process of breaking down organic matter into the mineral compo-

nents (mineralization) needed by living plants. Of the arthropods, termites, of course, play a key role in breaking down wood in rainforests throughout the world. In the rainforests of Sarawak (Malaysia), the abundance of termites varies according to soil type but ranges between 390 and 2270 insects per square meter. Termites alone can consume up to 16% of the dead plant material in some places. In African rainforests, the charming soil nests of the termite genus *Cubitermes* abound. Resembling towers of mushroom caps or little pagodas, they provide the termite hill and its inhabitants with shelter from the rain and consequent erosion. The African termites also decompose large quantities of soil and organic matter [41]. Although area counts produce fewer termites in Africa than elsewhere, there are more wood-lice (*Isopoda*), millipedes (*Myriapoda*) and beetles (*Coleoptera*). The result of these numerous organisms working diligently at the task of decomposition is a quick process of decay – less than a year. Leaves are most easily broken down, chunks of wood take longer depending on their size and density. The decay of organic matter in more temperate forests generally takes much longer than a year [42].

Complete Absorption of Nutrients

We do not yet know enough about tropical root systems to adequately understand how nutrients are absorbed by the plants. Biological aspects appear, however, to be more important than soil factors in determining the quantity and distribution of nutrients in the ecosystem [26]. An interesting example are rainforest trees – they often appear to have special relationships with organisms capable of absorbing and recycling nutrients. Small feeding roots can be found throughout the upper five centimeters and along the surface of the thin topsoil layer. But even larger roots seldom reach more than 30 centimeters into the soil. In a semi-deciduous rainforest in Ghana, more than 90% of the total weight of the root system is concentrated in the upper 30 centimeters [26]. Some tree species do have tap-roots and sinkers reaching two or three meters into the soil, but they are rather unusual. The root systems of even very large trees consist only of runners through the topsoil layer where they associate with fungi, particularly basidiomycetes and zygomycetes (terrestrial fungi). Fungus-covered roots are known as mycorrhizae ("fungus root"). Fungus growth covers the entire surface of the rootlets and may even penetrate them. This coating of fungal hyphae is connected with the mycelium spreading through the forest soil and serves as a second root system. It gives the forest tree the crucial advantage needed for the absorption of minerals. Thanks to mycorrhizae, practically no mineral nutrients are lost in a closed forest system. As soon as plant and animal material decays, the minerals released are returned to the system by the living plants' extremely efficient process of absorption.

The nitrogen cycle in a closed forest system is equally efficient. Not only is nitrogen absorbed by organisms in the soil such as bacteria (*Azotobacter, Clostridium* spp.), blue-green algae, fungi and actinomyces; but some higher species of plants are also capable of nitrogen fixation. Certain forms of algae and moss living as epiphytes on the leaves of higher plants fix nitrogen from the air and may pass it on to their hosts. In contrast to plantations, where the natural cycle has been interrupted, sufficient nitrogen is available for all the plants in a closed forest. Nutrients are absorbed so efficiently and so completely in the rainforest that plants can survive even without nutrients being stored in the soil. This is perhaps the most important difference between a tropical forest ecosystem and a forest ecosystem in the temperate zone. Trees have possibly developed even more unusual methods of getting the nutrients they need. A striking number of trees in the rainforest are hollow. Foresters consider such specimens to be "over-aged" but surprisingly enough, perfectly healthy trees are often discovered to be hollow when felled. The tree hollows are usually inhabited by entire colonies of bats or flying squirrels. The animals enter the tree either through an opening at the top or somewhere at the roots. Daniel H. Janzen who has done much research on the relationships developed between plants and animals in the rainforest, has pointed out that the animals' dung deposited at the base of the tree has a

Large-leaved *Marantaceae* and young raphia palms fill in a sunny spot on swampy ground.

The cabbage palm (*Antho-cleista* sp.) often occurs in gaps in the wettest rain-forests: here on the shore of the Sinoe River in southeastern Liberia.

high mineral content and acts as a natural fertilizer. So hollow trees may actually be an adaptation to poor soil [43].

Life in Eternal Twilight

On a clear day, sunlight filters down through the canopy scattering sun flecks across the forest floor. These bits of sunlight move with the changing position of the sun throughout the day and afford a stark contrast to the otherwise dim interior of the forest. The uppermost branches of the canopy receive the most sunlight which also allows the most intensive photosynthesis to take place here. Orchids also make their home in the upper stories of the forest. Much less light makes its way down to the lower levels. Less than 10% reaches the lowest canopy. On the ground below, light intensity measures only 1% or less of the light entering the tree crowns above, depending on the time of day. Although the human eye is capable of adapting to such twilight conditions, photography can pose a problem. With a film of normal speed, an exposure of half or one full second is needed to attain photographs with sufficient depth of

The leaf-ribs of the raphia palm *(Raphia hookeri)* (photo left) are used to fashion a variety of useful objects. The drawing above portrays a fish weir and was found in Büttikofer's "Reisebilder aus Liberia" published in 1890 [69]. Today in Liberia, one hundred years later, raphia weirs are still constructed according to this apparently well-suited model.

cecropia, is the most important pioneer species in many of Africa's moist forests. It springs up just about everywhere a gap has formed in the forest. Another common pioneer species throughout African rainforests is the cabbage palm *(Anthocleista nobilis).*

The dim light conditions near the forest floor make life difficult for all green plants at the lowest level. Even after a tree seed has managed to germinate, the seedling often has no future. Just barely existing as a small plant of 30 to 100 centimeters in height, it can only wait for a chance of a gap occurring in the canopy above which would allow it more light. The undergrowth of the rainforest consists for the most part of such tree seedlings and saplings having sprung up from the debris rotting on the floor merely to wait for the chance which may never occur. Only about 10% of the floor's surface is covered with ferns and herbaceous plants. They occur only where sufficient sunlight is present – in gaps and along the banks of rivers and streams. Some of the beautiful large-leaved Marantaceas, to which the South American arrowroot also belongs, are capable of existing under surpisingly poor light conditions, however. Grasses, on the other hand, are totally absent in a closed rainforest.

focus and a tripod is indispensable. The filter effect of a myriad of leaves allows not only green light but also a high proportion of long-waved red light to penetrate to the forest floor. The latter is apparently responsible for preventing pioneer species from germinating where the forest is intact. Their seeds do not germinate until the proportion of short-waved red light increases [26]. This is a sign for the dormant seeds that no canopy exists above and that new seedlings would have a good chance of filling in a gap. The umbrella tree *(Musanga cecropioides),* the counterpart of the South American

Serpentine roots anchor this tree on the shallow soil of the forest floor.

Among the many stunted tree seedlings on the dim floor of the forest, there are some which would never grow taller regardless of their share of sunlight. Dwarf trees often do not reach much more than a man's height. Where the soil is especially poor there may be several specimens of a poisonous dwarf beneath their tall neighbors. In Ghana, this species is known as "Kofie Kofie" which means nothing more than "go home, go home". This particular dwarf species (*Pycnocoma macrophylla*) has a crown of leaves up to 50 centimeters in length arranged in a rosette form. In closed West African forests,

about 17% of the higher plants (vascular plants) are woody plants which do not grow to more than eight meters in height (Tab. 6). Many of them are dwarf trees including dragon lilies (*Dracaena* spp.). A relative from Madagascar, (*D. marginata*) has become a popular house plant in many European homes where, similar to the rainforest, there is also often only dim light.

Trees – The Giants of the Forest

Middle to large sized trees make up about one third of the vascular plants in a West African

∨
A majestic liana twists its way up the buttressed trunk of this "African mango" (Irvingia gabonensis).

∨∨
Even the roots of giant trees reach only a few decimeters into the soil. The trees are supported by buttresses which, in the case of this kapok tree (Ceiba pentandra), may reach about 10 m up the trunk.

rainforest (Tab. 6). The crowns of the trees share the space available up to a height of 40 to 50 meters above ground and sometimes even higher. Just how they do this has led to much controversy among scientists. Originally, the rainforest was assumed to basically be structured in three layers. It has since become clear that the trees do not utilize the space according to a rigid pattern. Rather, they grow within their environment in such a way as to grant their leaves and branches the best possible position. This strategy may result in a rainforest of two layers, of four layers or of such an even distribu-

tion of tree crowns that distinct levels cannot be discerned. In addition, no one part of the forest is like the next. When an old tree of great size falls and creates a gap in the forest, new vegetation springs up in a process of continual change and restructuring over decades and centuries. It would therefore be incorrect to speak of definitive layers in the rainforest although certain areas of forest do show a clear structure of different crown canopies (Fig. pp. 64/65).

In the moist semi-deciduous rainforests of West Africa, the tallest trees rise to heights of 55 to 60 meters. Protruding well above the surround-

The fluted trunk of *Balanites wilsoniana*. The fruit of this species is a favorite of elephants.

ing trees, these giants are called emergents and give the top canopy of the rainforest its typically uneven appearance when viewed from above. Tree flora may be less diverse in Africa than in other rainforests, but it is no less impressive with great tree tops emerging like huge umbrellas held above their neighbors and measuring up to 30 meters wide. In Asian rainforests, more trees share the same crown area and the tree tops are therefore of lesser size.

Identifying Tree Species Can Be Puzzling

It is no easy task to identify tree species in the rainforest. Even a tropical forester can confuse closely related species and tropical hardwoods are often traded under the wrong name. Amateurs should not rely on the form of the leaves for they are often similar. Especially common are smooth-edged, leathery leaves with pointed drip tips allowing rainwater to run off more easily. Another problem is that one can often barely see into the trees above, making it difficult to decide whether these leaves or those leaves 30 meters up belong to the tree at the base of which one is standing. Foresters often cut a bit of bark from the tree to be identified since each species has a "slash" of somewhat different color and markings. A white, red or even light blue colored sap of milky consistency (latex) flows from the wound of many species. If doubt remains, the scent of the bark may give the final clue. Its surface, however, seldom gives a hint as to the species. Most trees in the rainforest have a cylindrical trunk with a smooth, thin, light-colored bark covering. Of course, there are exceptions: The medium-sized *Balanites wilsoniana* has a unique fluted trunk. Its fruit is a special favorite of elephants. Another unmistakable tree is an ebony which grows in the wettest rainforests. *Diospyros sanza-minika* has a very hard black bark full of lengthwise cracks. Elephants are also often attracted by this tree – as a

scratching brush. Some forest-dwelling people call it the "elephant comb". In the early days of African bureaucracy, the bark of *Fagara macrophylla* played an important role: Its thick, corky spines were used to make "rubber" stamps.

The shape of a tree's buttressed roots can also give an indication as to the species, though not always a reliable one. Buttressed trees are probably the most unusual structure in the lowest parts of the forest. They consist of both root and trunk. The kapok or silk-cotton tree (*Ceiba pentandra*) has buttresses of up to ten meters in height. On other trees, they form long, crooked runners along the forest floor. *Piptadeniastrum africanum*, a commercial species traded under the names dabema, dahoma and atui, has roots of this type. It has supple, feather-like leaves and belongs to the family of the Mimosacea. In the heat of the midday sun, its leaves close and allow more light to reach the forest below.

In earlier days, loggers had to build platforms some meters high around buttressed trees in order to get at the tree with their axe. Today, buttresses are quickly cut through with chainsaws. They clearly have a supportive function: Since the roots are generally quite shallow and therefore do not provide a strong anchoring for such large trees, a wide stance is necessary. There are places in West Africa where huge trees actually stand with almost no hold upon bare rock, held upright solely by the buttresses. On especially wet ground, some trees sit on stilt roots, an example being the "sugar plum" (*Uapaca guineensis*) of which the fruit is sold at local markets. The heavy stilts develop from thin adventitious roots which grow down out of the lower trunk until they finally reach the ground.

In swampy areas where the soil is poorly aerated, the roots of *Mitragyna ciliata* protrude diagonally and reenter the ground only to protrude again not far further. These knee-roots provide the root system in the swampy ground with additional oxygen. Raphia palms (*Raphia hookeri*)

In moist places within the distribution area of wet ever-green forests, *Cynometra ananta* often flushes young red leaves in April.

∨
Secondary vegetation along an old timber route in the well-protected Ankasa Wild-life Reserve in southern Ghana: Umbrella trees and young azobe (red ironwood, *Lophira alata*) with red leaves.

>>
Flushing leaves: limp, white leaves in the undergrowth.

are the most common trees found in swampy forest. Similar to mangroves, they also have aerial roots. Raphia swamps are found all over African rainforests where the ground is marshy and where water flows on the surface from time to time. Although raphia palms grow only 10 to 12 meters high, they form very dense thickets and are often not overshadowed by taller trees. Hikers in the rainforest do better to avoid raphia swamps since the microclimate there is much hotter and it is quite difficult to make headway. One should avoid raphia swamps unless, of course, one is looking for the palm trees. The raphia palm is a very useful plant. The young leaves are used to produce raffia fiber, which in turn is used to weave mats, hats and even textiles. The ribs of the leaves are used to build ladders and bridges and may also serve as scaffolding poles. When split, they can be woven into partitions. The ribs of the leaves alone have so many uses that raphia is locally called "the bamboo palm" [44]. There is also another reason why human footprints are not seldom found in the swamps: Raphia palms can be tapped. The sap is used to make a palm wine which is weaker but drier in taste than the wine made from the oil palm (*Elaeis guineensis*) also native to West Africa. Raphia wine is the poor man's wine to be had from the swamps of the rainforest.

Flushing Leaves

From brilliant emerald green to dull dark green, the rainforest offers every shade of color. But where are the red and violet orchids hanging in garlands from the trees and the dangling vines one expects to see in the jungle? Bright-colored flowers are rare in the undergrowth of the forest. Orchids are almost exclusively epiphytes and usually sit high in the tree tops. And even then, most African orchids are not spectacular with their greenish or yellowish coloring. But then suddenly that bit of red we have been looking for jumps into sight. It is not a flower we have spied, no, it is nothing more than a rosette of new leaves, brilliantly red-colored leaves. Sudden new growth of leaves is a common phenomenon in the rainforest. It is called "flushing" and that is about all scientists know of the process. As there are no distinct seasons in the wettest rainforests, one can only wonder what causes individual trees to suddenly produce thousands of new leaves. Is it temperature, light conditions or humidity that causes leaf flushing? Entire tree crowns turn red seemingly overnight. New leaves contain a high proportion of anthocyanine pigments which cause the red coloring. Not all quickly produced leaves are red however; one may also spy a cluster of shining white leaves in the twilight of the forest, hanging limply from a twig as though some hiker had found a hankerchief and left it on the nearest branch. These new leaves have only just begun to build up chlorophyll and their midrib and lateral veins will become rigid in a few days' time.

The Phenology of Rainforest Trees

It is difficult to determine what laws of nature regulate the production of new leaves, flowers and fruits in the rainforest. The seasonal changes – the phenology – of trees in a rainforest environment are so complex that one wonders if there is any pattern at all. Neighboring trees of one and the same species may be seen at different stages simultaneously: One tree may be "flushing" new leaves while another is bearing fruit and still a third has just started to flower. Sometimes even the very branches of the same tree appear to be experiencing different seasons. It is not unusual to find trees on which some branches are bearing fruit while others are energetically producing new leaves. Is absolute individualism the answer to the rainforest? Or are climatic factors not the only ones to play a role in the life cycle of a rainforest plant? Does the plant depend on the occurrence of specific insects for pollination? But what causes those insects to appear, and when? Many trees in the rainforest have adapted to pollination by insects with a long life span, insects which are present throughout the year – ants or beetles, for example. The flowers of such trees are situated on the trunk or even near the roots. These two unusual forms of flowering are called cauliflory and rhizoflory respectively and are an adaptation to pollination by non-flying insects – an adaptation, incidently, which is found in rainforests throughout the world. At least in regard to pollination, trees of this type do not depend on seasons and can afford a more individualistic behavior.

Fruit production also seems to be less dependent upon seasons. Different strategies have evolved: Some species almost continually bear fruit, others once or twice a year and still others once every few years. Trees belonging to the family of the dipterocarps common in Southeast Asia produce huge quantities of fruit quite suddenly and simultaneously after a long period of barrenness. The fruits are winged and spin away like tiny helicopters. The fact that these trees bear so much fruit at one time could be a defense against seed predators [35]: Instead of producing seeds evenly distributed throughout the year and risking their continuously being eaten by rodents before they can germinate, these trees have adapted by concentrating their seed production. Mass production increases each seed's chance to escape seed predators until germination takes place. There remains only the question of how the trees coordinate the simultaneous production of so much fruit! The situation is quite different for tree species which depend on particular animals such as birds or monkeys for seed dispersal. The fruits of these trees must be eaten since the seeds need to pass through an animal's intestine or at least

be dropped at a distance from the seed bearing tree in order to germinate. In this case, the simultaneous production of large amounts of fruit would have just the opposite effect: A surplus of fruit would reduce the chance of reproduction. It is more advantageous for trees of this type to bear smaller amounts of fruit at intervals throughout the year.

Even in West Africa's seasonal rainforests, there is no strictly seasonal production of fruit. Species with winged seeds also appear to adopt the strategy of simultaneous seed production whereas species dependent on animals to disperse their seeds bear in an uncoordinated fashion throughout the year or simultaneously but at shorter intervals (Fig. p. 85). The kapok tree (*Ceiba pentandra*) has a very seasonal phase of seed production. Its seeds must be ready during the dry season. They are embedded in a floss of fine whitish hairs (the kapok) easily blown away by the wind. During the wet season, seeds of this type would be less likely to be well dispersed. But even these observations give us only very rudimentary knowledge of how trees reproduce. Phenological observations would have to be made over many years in order to solve the puzzle of seed dispersal in the rainforest. Yet there may be no simple answer to how the life cycle of a rainforest tree functions. In spite of all the scientific interest and human tendency to analyze and order what we know, it may be just as well to simply admire the wonders of the rainforest. The various and endlessly diverse interrelationships within the forest are in any case much more complex than we could ever imagine.

Adept Climbers

Not without reason have creeping and climbing vines become the trademark of the rainforest. Where there is little light and a tough competition for a place in the sun, one may expect plants to specialize. About one quarter of all

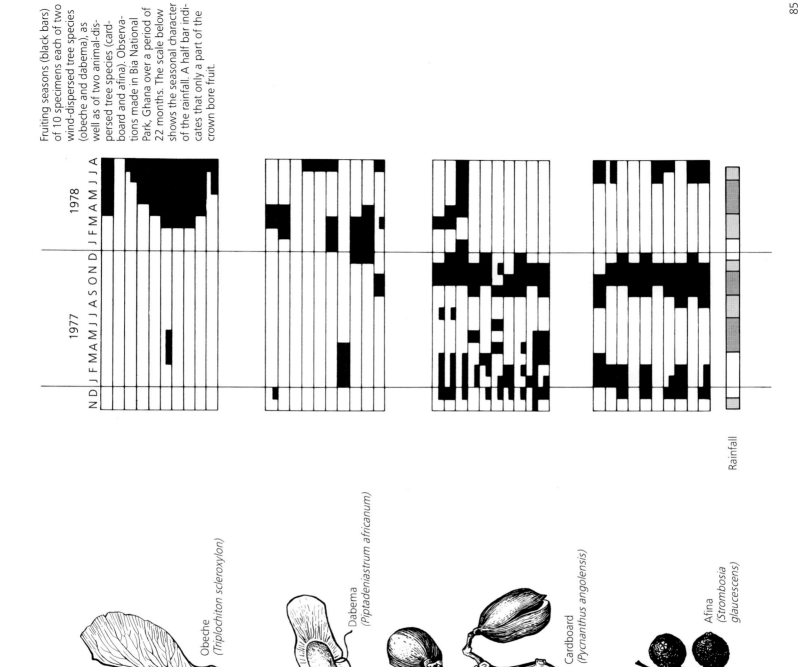

Fruiting seasons (black bars) of 10 specimens each of two wind-dispersed tree species (obeche and dabema), as well as of two animal-dispersed tree species (cardboard and afina). Observations made in Bia National Park, Ghana over a period of 22 months. The scale below shows the seasonal character of the rainfall. A half bar indicates that only a part of the crown bore fruit.

1978

1977

N D J F M A M J J A S O N D J F M A M J J A

Rainfall

Obeche
(Triplochiton scleroxylon)

Dabema
(Piptadeniastrum africanum)

Cardboard
(Pycnanthus angolensis)

Afina
(Strombosia glaucescens)

woody plant species in a West African rainforest are large lianas. And there are also quite a number of smaller climbers. They flee the twilight below by entwining their way up trees and often reach the uppermost branches. In West Africa, we find some of the largest lianas. Those of the genus *Strychnos* may grow to a base diameter of 30 centimeters. The sea bean (*Entada pursaetha*) finds its way into the top canopy with vines 15 centimeters thick and produces pods over one meter long. The large, shiny brown seeds are called Queensland beans in India and Australia, where they are roasted and eaten. The seeds are capable of germinating even after long periods of immersion in water. They are occasionally found along the shores of rivers or on beaches [44]. It is therefore not surprising that this impressive liana is one of the few rainforest plants which also occurs on other continents. The sea bean is found in moist tropical Africa, Asia and in northern Australia. Lianas share a common fate with the trees they have entwined themselves with: Should a tree fall, the liana will go down with it. Once a gap has opened in the forest, small herbaceous creepers take advantage of the extra sunlight and quickly begin to cover the fallen trunks. Secondary vegetation continues to fill in the gap and soon fast-growing lianas with woody stems are making their way up to the sun. Lianas are not necessarily restricted to one tree. They may be entwined among several trunks and reach a respectable age together with their hosts. Sometimes, one tree will support a number of different lianas, which in turn are themselves entwined. The tangles of vines provide living space for monkeys, squirrels and other animals high above the ground. Should a supporting branch fall, loops of lianas will hang down between the trees – a classic picture captured by many a photographer in the jungle. Lianas often survive even the fall of their supportive tree. New shoots find their way up the nearest neigh-

bor. Tangles of lianas clutter the forest floor where a tree has fallen. Foresters are not especially fond of lianas as they compete with the trees for light in the top canopy and tie them up with each other. When a tree is felled, the thick lianas ripped down with it can become dangerous whips. A logger will therefore first tend to the lianas encircling the tree of his choice before cutting into the trunk.

Not all lianas are climbers from the start. Some begin life as common bushes or small trees. Only after some years do they begin to develop into vines and become increasingly thicker. This kind of liana can come to the aid of a thirsty traveller. The sap of *Tetracera potatoria* is a clear watery fluid suitable for drinking and flows abundantly from the liana when cut. But unless one is sure of the species, it would be better to wait for the next stream. There are nearly 100 species of liana in a small area of West African forest. The sap of some of them serves as antiseptic for wounds, bird-lime and arrow poison!

Humus in the Tree Tops and the Stranglers of the Forest

A varied group of plants has managed to entirely free themselves from the ground in their strive for sunlight: Epiphytes. The high humidity in the forest allows them to grow on other plants, mainly trees. Far more than half the trees in a closed rainforest are inhabited by epiphytes. A study in undisturbed areas of rainforest in Liberia showed 153 different species of epiphytic vascular plants growing on 47 species of tree. 66% of the epiphytes counted were species of orchid, 25% were ferns and the remaining 9% were other plants [45]. Of course, species of lower plants, algae, lichen and moss also live as epiphytes and may even be found on the surface of leaves. This type of epiphyte is called epiphyll. The variety of epiphytes is lower in

The spreading root network of a strangler fig (*Ficus* sp.). It begins life as a modest epiphyte at the base of a branch on the tree's trunk.

opened up secondary forest. Lianas, too, are fewer and less varied. The broken canopy above causes the microclimate below to become drier and less hospitable for plants depending on high humidity.

Epiphytes do not live on air and water alone. Much of the humus produced by the decay of leaves, twigs, flowers and fruits in the rainforest is found high above ground. Humus develops in tree hollows, between forked branches and on the upper side of large branches [26]. In addition, small particles of soil are carried up from the ground by ants and termites. Old trees are often covered with entire gardens of epiphytes. In some of the more hilly regions of Ghana where evergreen upland rainforest dominates, up to 40 species of vascular epiphytes were found on individual large trees [28]. Some tree species themselves absorb nutrients from the little pockets of humus with aid of fine adventitious roots. Although strictly speaking, epiphytes grow on and not from other plants, they are not always as benign as one might think. The roots of epiphytes are often associated with fungi, the mycelium of which may penetrate the tree and absorb nutrients from its vascular system [35].

Thus, mycorrhizae indirectly lead some epiphytes to become parasites.

The stranglers of the forest are also epiphytes. They usually begin life sprouting in the fork of a branch from a seed deposited with dung left by a bird or a monkey. The seemingly innocent epiphyte sends an ever longer and wider web of roots down the trunk of its host until it reaches the ground. And then the tragedy begins: Once the roots begin to absorb nutrients from the soil they thicken and entwine themselves more closely around the host. The once modest plant becomes a tree of its own spreading its branches above those of its victim. The host tree is deprived of more and more light and its vascular system becomes ever more constrained. The roots of the strangler close around the poor tree, which dies and decays. In the end, the strangler stands alone. All that remains to remind us of the strangled host is the hollow within its trunk. Gaps between the former roots of the strangler show us where it once stood. West African stranglers are all species of fig trees (*Ficus* spp.) and are not easy to identify, as is often the case with criminals.

Little-known Rainforest Fauna

The mother of this young royal antelope fell victim to an Akan hunter. The delicate meat of Africa's smallest hoofed animal – adult specimens reach a shoulder height of only 25 cm – is reserved for the village chief.

If ever a Goliath beetle fell from the tree tops to the red sand of the village below, it would stir up quite a bit of surprise among the inhabitants. Chances are it would be their first encounter with the giant beetle. Although the villagers are familiar with the many animals they hunt, they have probably never seen this strikingly marked black and white beetle living only 40 or 50 meters above in the tree tops at the edge of their village. Like many tree-dwelling species, the Goliath beetle is scarcely known to the jungle's human inhabitants although they regularly hunt and gather in the rainforest. But scientists have even less chance of getting acquainted with such hidden species. It is the visibility of a species which finally determines the knowledge of its occurrence, ecology and sociology. The libraries are full of scientific literature on East African large game – easily observable subjects. But what scientist would risk doing an ecological study on the African linsang (*Poiana richardsoni*), a viverrid which lives and hunts prey in the uppermost branches of the rainforest canopy?

Whoever spends their first night in the rainforest does not doubt that the jungle is full of mysteries and much still to be discovered. An orchestra of calls fills the night – from schrill screeches to tender cooing – and one quickly forgets that no large animals were sighted during the day. Not even the experts can identify each call used by the noctural creatures of the forest to communicate with other members of their species. In European forests, apart from some exceptions, birds alone communicate by calls. In the rainforest, acoustic communication is an absolute necessity for mammals as well. Vegetation and noctural activity make visual contact impossible. The fact that many rainforest mammals spend their lives high above ground amongst the trees and often lead solitary lives further hinders communication. Since so many species share the same habitat, the population densities of individual species are generally low. This means that individuals of the same species are often at a distance from each other. But especially for smaller mammals, adequate estimates of population density have hardly been made. Empirical methods used in closed forest moreover tend to result in underestimations.

Probably the strangest sound in the night is a succession of penetrating screams which follow each other with increasing rapidity, becoming louder and louder to finally culminate in a screech of panic. The drama begins shortly after twilight and newcomers to the forest wonder what frightening tragedy is taking place behind the green curtain. Any contemplation, however, is soon interrupted by the next sequence of screams in the distance followed by a third and a

A large part of the rainforest fauna spends its life high above ground : Freshly-cut vertical profile through a moist semi-deciduous forest in Ghana.

fourth. The unnerving sounds continue until dawn and always seem to eminate from the same directions. After some time, one may realize that certain animals are apparently marking their territory acoustically. What may be even more of a surprise is that the species so inclined is a close relative of the elephant and no bigger than a rabbit. The tree hyrax is seldom seen by man except when it is unfortunate enough to have nested in a tree chosen for felling.

Variety in the Tree Tops

The fauna of African rainforests is rich in smaller mammal species seldom seen by man. Most are nocturnal and live exclusively in the trees. Arboreal mammals, however, are not limited to Africa. 45% of the mammals in Borneo live high above the ground, bats and squirrels not included. The percentage in Africa is lower, about 30%, due to the generally higher proportion of ungulates, mainly antelopes, in African rainforests. Many species of mice crawl up trees but the only known exclusively tree-dwelling mice belong to the genus *Thamnomys*. They climb to the highest branches of the canopy where they are often preyed upon by the African linsang [46]. With regard to other rainforests, comparatively fewer African mammals possess physical adaptations to life in the trees. Aside from certain monkeys, the tree pangolin and the long-tailed pangolin with their long prehensile tails; two lemur species, potto and dwarf galago, with adapted extremities; and four species of flying squirrels able to glide from tree to tree. One African reptile has also discovered flight: Günther's spiny lizard (*Holaspis guentheri*) can flatten its body to such an extent that it is able to glide from one trunk to the next [47]. Although numerous squirrel species, certain viverrids and species of mice are also almost exclusively tree-dwelling, they possess no special physical adaptations. Even the tree hyrax, which hardly ever sets foot on earth, can scarcely be distinguished from its ground-living relative the rock hyrax.

Arboreal mammals, reptiles and amphibians are less notable in regard to their species diversity and possibly even their biomass than the arthropods which are found throughout the forest from the ground to the tops of the highest trees. Among the arthropods, the class of insects has especially presented researchers with a number of surprises during the last few years. Only today is the sheer quantity of different species inhabiting the upper canopy of rainforests becoming clear and it surpasses our wildest imaginations.

The Worldwatch Institute in Washington cites known plant and animal species throughout the world in 1987 at 1 390 992. Just over one million of those, or 75%, are animal species [48]. In a publication of the same date, the World Conservation Monitoring Centre (WCMC) in Cambridge cited 1.2 million species of arthropods alone. The WCMC quotes the total number of scientifically recognized species at 1.8 million, although the figures listed appear to be rounded up [49]. Nevertheless, one seems to agree that the species known today are a mere fraction of those which actually exist on earth. The reason lies in our ignorance of untold numbers of insect species in tropical rainforests around the world, not to mention the mites.

The higher one goes in the zoological system, from lower animals to the most highly developed mammals, the more can be said about their species diversity and ecology. This fact does not pertain to the rainforest alone but there it is even more so. The taxonomic classification of lower animal species in the rainforest has only just begun. Since vertebrates have been most extensively studied, the chances of discovering new species diminish as one moves from fish to amphibians and reptiles, to birds and finally, to mammals. But taxonomic classification is just the first step to understanding the extremely complex ecology of species and their interrelationships which make up the rainforest.

No small animal is safe from driver ants – small comfort for the grasshopper being eaten alive.

The architects of these pagodas are of the termite genus *Cubitermes*. Their soil nests withstand heavy rainfall and rainwater dripping down from the canopy above.

With a few driver ant soldiers in his pants, any hiker will make a run for it. Admiring the wonders of the rainforest has its price.

Astounding Insect World

The only insect family which has been more or less thoroughly classified is the swallowtail family (*Papilionidae*). Not surprisingly, they include some of the largest and most attractive butterflies – swallowtails, swordtails, jays and apollos. West Africa harbors 32 of the 573 species in this family [50]. Ants, too, have been quite extensively classified in Africa. Compared to other insect groups, there are fewer species of ants but they can make themselves all the more noticed by their numbers: Hikers in the rainforest are alerted to their presence by whomever is leading the group – when that person begins to run fast it is an unmistakable sign that driver ants are near. There are 50 to 60 species of driver ants in Africa, most of them belong to the genus *Anomma*. Whoever has the ill luck to meet with an army of driver ants fanned out and spread through the leaves and branches of the forest will rapidly retreat. In the face of a million ants, he or she can only flee to safe ground and do a quick striptease to get rid of the tenacious creatures already clinging to the skin. But the sharp jaws of the workers and particularly the soldiers do not do serious harm to man. The attack was only error, the ants' raiding technique is an adaptation to their usual prey: millipedes, spiders, social insects and other arthropods. After periods of rest, driver ants begin to migrate during the rainy season for 12 to 18 days searching for nests of possible prey populations [51]. An entire army of driver ants, often over one and a half million strong, will move in long columns across the forest floor and through underground cavities. Driver ants are also a constant danger for

smaller mammals and reptiles. During the rainy season, they are the python's worst enemy. Many Ashanti hunters in Ghana are convinced that after a python has strangled a larger animal, it makes the rounds of the forest before devouring its meal. The snake must assure itself that there are no driver ants nearby for it would not be likely to escape weighed down with such a heavy meal [52].

By far, the most diverse order of insects in the rainforest are beetles. In recent years, confronted with the continual clearance of tropical forest, the scientific community has begun to study the possible consequences for species diversity. Especially American scientists have made some surprising discoveries: Terry Erwin of the Smithsonian Institution used Pyrethrum, a biodegradable insecticide, to fog a particular species of tree (*Luehea seemannii*) in Panama and collected 1200 species of insects which fell from its branches. He also found 41 000 species of arthropods in a single hectare of Peruvian rainforest, more than a quarter of which belonged to the beetle family. That already equals the number of beetle species in all of Central Europe. In Amazonian rainforests, Erwin found many beetle species to be dependent on a particular type of forest. According to new extrapolations resulting from Erwin's studies, there could be up to 30 million insect species throughout the world [53,54]. That would mean there are 34 species still to be discovered for each one known today.

In 1985, British and Indonesian entomologists conducted an intensive investigation of insect fauna in Dumoga-Bone National Park on the island of Sulawesi. They, too, employed the fogging technique and collected the insects which fell into the tarpaulins set up below. The proportion of known to unknown species was calculated according to the catch. Further investigations are still needed to confirm Erwin's extrapolations of total species diversity but even prudent scien-

tists assume there are at the very least ten mil-lion species of plants and animals on earth. An overwhelming majority of those species are insects hidden away in tropical rainforests. We know nearly nothing about more than 99% of them, whether they have been identified and described or not. For every 100 000 supposed insect species, there is but one ecological stu-dy [55], a situation which is not likely to change in the next few years. Taxonomists do not have the financial means to even inventarize the immense number of species, let alone carry out more detailed studies. Entomological research today is concentrated mainly on disease carriers and agricultural pests. For now, we must there-fore live with the fact that every major clearance of tropical forest probably wipes out innumera-ble species never identified. Peter Raven, director of the Missouri Botanical Garden, believes that several species of plants and ani-mals are already being lost daily. Naturally, spe-cies loss does not only affect the actual victims but also those species of plants and animals with which they were somehow interdepen-dent. If forest destruction continues at its pres-ent pace, the rate of extinction could increase within the next 20 to 30 years to several hundred species a day [48]. It is, however, nearly impossible to confirm estimates of this kind since most of the extinct species have never been recorded. When one considers the limited distribution area of most rainforest plants and animals and the fact that many insects are re-stricted to a particular species of tree however, Raven's conjecture does not seem far-fetched. The recent studies on species diversity and rates of extinction are especially important for the rainforests of West Africa. Today, African rain-forests are generally less researched than those in Latin America or Southeast Asia and yet in West Africa, they are much more fragmented. It is thus quite probable that species extinction, at least among insect species, has already begun.

Currently, only Africa's largest butterfly (*Papilio antimachus*) is listed as rare although it occurs in just about all of the continent's rainforests [50]. Its wing span can reach up to 23 centimeters. The privilege of being the only insect species listed as rare in Africa is due to the fact that the giant butterfly can hardly be overlooked and that it belongs to the best studied insect family on the continent. But the Upper Guinea forest block alone harbors some 750 butterfly species and in all of West Africa, including the savannah zone, there are probably 20000 species or more of moths [56]. Most of them are assumed to have very limited ranges since they are usually dependent on a specific plant for their develop-ment. As we will see with regard to higher ani-mal species as well, the distribution areas of many West African animals are relatively small which makes them extremely vulnerable to for-est destruction.

Fish on the Ground, Frogs in the Trees

Interestingly enough, the scientific classification of fresh water fish is far from adequate although they provide a substantial proportion of the diet of many forest-dwelling peoples. Not only has little research been done on the fish fauna in the Amazon Basin – where a very high percentage of the species existing have probably not yet been recorded – but there are also many un-known species in West and Central Africa. Such statements, however, do not mean that these fish have never been known to man: Forest-dwelling people and local populations through-out the tropical forests of the world have long made use of "unknown plant and animal spe-cies". If asked, forest people could certainly enlighten scientists as to the local name and the biology of the one or the other species. West African fish have only been systematically stud-ied in the waterways of the northern savan-nah [57]. But recently, a fish study was also

< Earthworms of the family *Megascolecidae* reach well over one meter in length. But they play a lesser role in the decay of organic material in the rainforest than do earthworms in the temperate zones of the globe.

<< These boys know what they have fished out of the Bia River in western Ghana – scientists may not be sure. There is as yet no comprehensive description of the fish fauna of African rainforests.

made in and around the newly established Korup National Park in the rainforests of Southwest Cameroon [39]. Korup borders on the brackish water zone of the Rio del Rey in the south with its large stand of mangroves. It also lies within the "West Central" center of endemism located between Cameroon and Gabon. 140 different species of fish were found in the waters of this rather limited rainforest area, a number which surpasses that known in larger African rivers such as the Nile or the Zambesi River. Several of the species had been known to exist only in the Congo Basin and ten were completely new.

Among the vertebrates found in the rainforest, whether in South America or in Africa, fish probably hold the most secrets yet to be discovered, if ever.

The Korup study led the researchers to believe that many fish species spawn before the wettest season of the year. Once the heavy rains have flooded parts of the forest, the young fish leave the streams and swim along the forest floor where they find fallen fruits and drowned insects. The rivers and streams alone would not hold sufficient nutrients for them to grow. Following the heavy rains, the offspring are past the critical stage of their development and return to the waters as the flood recedes.

Thirty years ago, neither did we have much idea of the amphibians in West African rainforests. Several species are tree-dwelling, difficult to find and not easy to identify. Then in the 1960s, a Danish zoologist, Arne Schiøtz, began a thorough study of amphibians in West Africa: In Nigeria for example, he found 78 species, 47 of which had never before been recorded in that country [58]. Schiøtz was an especially avid fan of frogs and of tree frogs (Rhacophoridae) in particular. These graceful creatures of beautiful coloration spend most of their time in the branches of bushes and trees. Their toes are equipped with adhesive disks. They lay their eggs in foam nests hung between the branches or in tall grasses above the water. In Africa, tree frogs occupy the ecological niche in the rainforest filled by some arrow poison frogs (Dendrobatinae) in South America. An astounding array of tree frogs is also found in the African savannah. An incredible variability in the dorsal coloration of some species has led to a chaotic nomenclature. Schiøtz collected a total of 4200 specimens of tree frog throughout West Africa, from Sierra Leone to Cameroon and managed to put the zoological systematics into order. Of the 54 species he described in detail, 18 were new. But his most important discovery concerned the distribution of the different species [22]: For the most part, Schiøtz made his rounds of the forest and savannah at night and during the mating season whereby he learned to identify and track down different species by their calls. He found that there were always 10 to 11 species of tree frogs in different areas of closed forest but they were not the same ones everywhere! The species gradually change from west to east and only three occur on both sides of the Dahomey Gap. Species which geographically replace others are called vicariating species. So Schiøtz discovered that every bit of rainforest has its own community of 10 to 11 species of tree frogs belonging to a total of 23 different species found between Sierra Leone and Cameroon.

Up to now we have spoken only of tree frogs occurring exclusively in closed forest. Altogether different species live in secondary vegetation and on fallow land. In one specific area, 5 species of tree frogs are likely to occur and throughout West Africa there are 12 vicariating species adapted to secondary vegetation. The situation is similar again in the savannah but of course with different species. A savannah community will hold 7 to 9 species depending on the area. A total of only 10 species occurs in savannah habitat but they have a wider range and it is more difficult to discern just when one species has replaced the next. West African tree frog fauna thus consists of clearly separate species groups distinguished between forest, secondary vegetation and savannah habitat. In addition, there are a few species adapted to special habitats such as upland forests. Thus, every area of West Africa has its own unique community of tree frogs.

With his countless nightly expeditions in West Africa, Schiøtz discovered certain patterns of distribution among closely related species as there are within other groups of animals as well. But few researchers have ever been as scrupulously exact. It is not by chance that the 12 tree frog species found in secondary vegetation are those most commonly found in museum collections.

Until Arne Schiøtz undertook his studies, those were the species considered representative for tree frog fauna throughout the rainforest. But the specimens swimming in formaldehyde on museum shelves were merely the random catch of collecters who never set foot beyond the bounds of the red sandy roads winding through the forest. Until recently, our picture of the rainforest generally corresponded to that of its secondary vegetation, of the underbrush along the paths and edges of cleared land.

The Frightful Reptiles of the Forest

Like amphibians, reptiles are also easier to observe on fallow land and particularly on farmland – snakes, for example. Here mice and other rodents are attracted by crops and in turn are followed by their enemies. There, on farmland, one may run into an unpleasant visitor. About 100 deaths from snake bites are recorded annually in Ghana, especially in isolated farming areas far from medical care [52]. Potentially deadly snakes are not rare in West Africa. The bite of a Gabon viper, rhinoceros viper, cobras, green mamba and a few other snakes will certainly have serious or even fatal consequences. One seldom runs into these snakes in closed rainforest, however, and should one be sighted, it is often impossible to identify since any snake will quickly retreat into the darkness of the undergrowth.

It is possible that the density of snakes in the rainforest is higher than one may suppose. From holes in the ground, deserted termite nests, and hollow tree trunks to the branches high above, there are numerous hiding places for snakes to escape our view. Over 100 snake species occur

in West Africa with representatives from all major groups except rattlesnakes and sea snakes. Most of these species are found in the rainforest. But scientists know little of their geographical distribution and absolutely nothing about their abundance in a specific habitat. Whoever shys the rainforest for fear of snakes may be assured to know that it would be pure luck to even glimpse a snake in West Africa, worked dale, an expert on snakes in West Africa, worked for 14 years in Ghana's forest service which employs hundreds of people in all parts of the country's forests every day. Only two cases of

snake bites occurred during that time and neither was fatal [52].

West African rainforests harbor both the largest and the smallest snakes on earth. The African or rock python *(Python sebae)* is common in many places south of the Sahara. By far the largest specimen ever found came from the rainforests near Bingerville in the Côte d'Ivoire. The unusually large python measured 9.96 meters in length [52]. In spite of rumors, not even the anaconda *(Eunectes murinus)* in South American rainforests grows to a greater size. The other West African python species, the royal python

101

Three crocodile species live in the rainforests of West Africa: the dwarf crocodile (left), the Nile crocodile (middle) and the slender-snouted crocodile (right). Of the three species, the Nile crocodile with a length of up to about 7 m grows to much larger size than the other two species.

(*Python regius*), attains a length of only 1.5 meters and is not a typical inhabitant of closed rainforest. Then there is the Calabar ground python (*Calabaria reinhardtii*) which grows to about one meter in length and searches for worms on the forest floor from Liberia to the Congo Basin. The worm snakes (*Leptotyphlopidae*) of West Africa are among the smallest snakes known: They grow to only 15 centimeters in length and their diameter is scarcely that of an earthworm. There are about eight of these miniature species of snakes in West Africa and all are exclusively terrestial. Nothing is known

about their distribution however. Worthy of mention are perhaps two other species of likewise harmless water snakes which are quite common in certain places. They occur in both stagnant water and forest streams where their diet consists of fish and frogs: Smyth's water snake (*Grayia smythii*) and the brown water snake (*Natrix anoscopus*). A brown water snake has occasionally frightened a fisherman upon emptying his trap woven of raphia palm leaves. Only a trained eye will spy still another reptile in the dim light of the forest: the West African hinged tortoise (*Kinixys erosa*). Withdrawn into

Crocodile skins are tanned in the backyard of a poorer neighborhood in Abidjan – even today!

its shell, sitting motionless beneath a piece of rotting wood, this up to 30 centimeter long tortoise is easily passed by without notice. But once it raises the hinged rear part of its shell and stretches out its legs to race across the ground with surprising speed, one is sure to spot it. The tortoise is commonly captured by the local forest inhabitants.

West African crocodiles also lead rather secretive lives. The slender-snouted crocodile (Crocodilus cataphractus) lives in forest streams and rivers in the shadows of the rainforest. It is a shy creature and may be best observed at night with a flashlight and a bit of luck. The German zoologist Ekki Waitkuwait invested much time, energy and patience over the years observing crocodiles in the rainforests of the Côte d'Ivoire. He especially studied the breeding biology of the slender-snouted crocodile [59,60]. Like other crocodile species, this up to four meter long species takes particular care of its offspring: At the beginning of the rainiest period from March to April, the female builds a nest of dead leaves and other plant material near the shore of a forest stream where she buries 10 to 20 or more eggs. After 90 to 100 days, when rainwater has filled all possible channels in the forest, the young begin to hatch. The offspring cry out loud enough to attract the mother crocodile, who digs out the hatching young and frees them from the nest.

The considerably smaller West African dwarf crocodile (Osteolaemus tetraspis) is less dependent on open water for its habitat. It lives in damp areas and is often found in the raphia palm swamps of the wettest rainforest types. It is not common in the drier types of forest. The dwarf crocodile moves with surprising speed through the often spiny undergrowth of the swamp forest and also builds nests of dead leaves, smaller but similar to those of the slender-snouted crocodile. During his studies in the northeastern part of Taï National Park, Waitkuwait even found a skull of an adult Nile croco-

dile (Crocodilus niloticus). Nile crocodiles do occur in West African rainforests but when inland they usually keep to the sunny, sandy beaches of large rivers. More frequently, they are still found in the lagoons along the West African coast [60]. Although hunting this reptile is illegal, poachers often stalk the Nile crocodile for its skin.

A Random Selection of Birds

As one might expect, tropical rainforests harbor an array of bird species. The varied levels of the forest, together with the diversity of fruits and insects provide a wonderfully suited environment and numerous ecological niches. Indeed, no other kind of habitat harbors as many bird species as the rainforest. The number of bird species inhabiting a rainforest, however, is far from constant but varies in space and time. Bird fauna in swamp forest, for example, greatly differs from that in lowland rainforest or in mountainous regions. In addition, bird species also have a limited distribution often defined by climatic events long past. Species dependent on the lower levels of primary forest disappear in areas where too much secondary vegetation has grown, particularly in areas where farming activity has disturbed the forest. Species which otherwise occur only in open forest take over the altered habitat. And then there are species that occur in both open and closed forest as well as others which change habitats seasonally. Neither is much known about just which European migratory birds spend the cold months in African rainforests. Clearly, it is not easy to judge the diversity of bird species in a specific area of rainforest. In the case of any estimate, details on where the observations were made must be included as well as whether the sightings were made in secondary forest, plantation areas, whether open water was near or if the area of

observation can be considered closed and un-disturbed rainforest.

The result of a bird survey is also influenced by another factor – the ornithologist's persever-ance. The Koepckes, a husband and wife team, spent two years watching birds in a 2.5 square kilometer area of lowland rainforest in eastern Amazonian Peru. They counted 320 species of which 210 were true closed rainforest species. John O'Neill from Louisiana State University made observations in a comparatively small area of rainforest in the same Peruvian region. O'Neill repeatedly spent a number of months bird-watching over a period of eight years and observed 408 species, a surprisingly high number. About 300 of the species observed were closed forest birds. Besides the high number of species, it is also interesting to note that O'Neill's list of species continually grew with each additional period of observation and he

believed there were many more species to be discovered. A number of the birds sighted were quite rare for the region and were seen only once [cit. in 61]. Needless to say, comparisons of ornithological species diversity between differ-ent rainforest areas must be made with extreme care. Days and even weeks of observation are obviously not enough to establish a reliable checklist. Even studies of longer duration are not necessarily accurate. There is no doubt, how-ever, that there are generally fewer bird species in a comparable area of African rainforest than in South America and, incidently, also fewer bat species. Just why the diversity of flying verte-brates in Africa is relatively limited is uncertain. A possible loss of species during the Pleistocene when the rainforests had shrunken to small re-fuges can scarcely be the cause. Similar cool, dry periods led to forest refuges in South America as well.

This photograph from Büttikofer's "Reisebilder aus Liberia" (1890) shows the arrival of slain pygmy hippopotamus in a forest village. The man in the foreground is carrying a banded duiker. The hunting party is welcomed with horn blows from a bongo horn. Such horns were also used to alert the villagers to danger and attacks by enemy tribes.

Bird species in African rainforests have been relatively well documented. Comprehensive studies have been published on birds in the Congo [62] as well as in West and Central Africa [63]. It would be quite unusual today for an ornithologist to record a species never seen before. 266 true forest species of birds occur in the Central African rainforest block and 182 species in the Upper Guinea forest block. Many of those species, of course, occur in both areas [64]. These numbers are likely to reflect the diversity of the true forest bird fauna. Estimates of overall species diversity, on the other hand, are not very helpful in judging local species diversity, for example within a specific square kilometer or within a national park. As we know, certain bird species have a very local distribution and do not occur throughout a rainforest block. Relatively adequate maps of distribution ranges exist only for passerine birds [18].

To further complicate matters, many birds not typically found in the rainforest actually appear here and there causing some confusion in any survey. There are very few species lists of birds in clearly defined areas in West Africa. Most surveys were not carried out for sufficiently long periods. To date, no comprehensive survey has been made of birds in the most important protected forest area in the Upper Guinea forest block, Taï National Park in the western Côte d'Ivoire. A survey in the Gola Reserves, an area of about 750 square kilometers in the southeastern corner of Sierra Leone, recorded 142 bird species including species on open water and in disturbed forest [65]. In the Bia Reserves in Ghana, a total of more than 300 square kilometers, 160 species were sighted over a period of three years. Here, too, a number of atypical species were included in the count [66]. But even these figures are very likely incomplete. A com-

A rare photograph of a wild pygmy hippopotamus. Due to its limited area of distribution in West Africa, the pygmy hippo itself must be considered rare and endangered.

parable area of primary forest in the Upper Guinea forest block lacking open water surface probably holds about 200 species of birds. Even higher numbers can be expected at the eastern end of the West African region, in the border forests between Nigeria and Cameroon. The combined influence of the two centers of endemism, the "West" and "West Central" endemism, the "West" and "West Central" (Fig. p.39), has led to an especially high diversity there. 252 species of birds have been recorded in and around Korup National Park, an area of 1200 square kilometers in West Cameroon. 129 of the species observed were passerine spe-

cies [67]. Although the count is most probably incomplete, it may be one of the highest figures recorded for a specific African locality.

Discovering West African Mammals

There are far fewer questions as to the species diversity of mammals, animals often hunted and relatively easy to observe. But even then, new species of rodents or bats may still be discovered and there may be new records of smaller mammals for West Africa if sufficiently large forest areas remain. Even within the Upper

Guinea forest block, from the most western reaches of the rainforest in Sierra Leone to the Dahomey Gap, the array of mammal species changes. In Liberia and the western Côte d'Ivoire, the "West" center of endemism, species diversity is somewhat higher. Those rainforests correspond to the forest refuge area which existed during the Pleistocene and where plant and animal species survived the cool, dry climate. Today still, the area harbors more species than elsewhere.

The first zoological collections from West Africa were taken from this area in the mid-nineteenth century: In 1840, Samuel Morton sent specimens of animal species from Liberia to the Academy of Natural Sciences in Philadelphia. Among them were skeletal parts of a pygmy hippopotamus thus earning the creature its scientific name *(Choeropsis liberiensis)*. But it was not until 1912 that Hans Schomburgk managed to capture and export three living specimens of this rare and mysterious animal to New York [68]. The pygmy hippopotamus is endemic to West Africa and occurs only between Sierra Leone and the Bandama River in the Côte d'Ivoire. The pygmy hippopotamus, a usually solitary animal which may also live in pairs, has never been closely studied. Only in the early 1980s did the species become listed as endangered. Until 1930, a second population existed on the Niger Delta 1300 kilometers east of its major area of distribution. The Niger population was probably isolated thousands of years ago by the reopening of the Dahomey Gap following a warmer moist period.

One of the earliest reports on animals in West Africa is in Johann Büttikofer's *Reisebilder aus Liberia*, published in 1890 [69]. Büttikofer was born in the small Swiss village of Ranflüh in Emmental where he later taught school. After learning taxidermy on his own, he left Ranflüh to work as a taxidermist at the Museum of Natural History in Berne. In 1878, he moved on to the

Museum of Natural History in Leiden (Holland). One year later, he went on a private expedition to Liberia accompanied by Carl F. Sala who died in Robertsport in 1880. Büttikofer returned in 1882 and sold his entire collection to the Museum where he was re-employed in 1883. It was not long before he again set off for the "Fever Coast". Accompanied by F.X.Stampfli in 1886, Büttikofer explored the rainforests in Liberia on foot and by canoe. He was able to complete much of his collection. Stampfli had undertaken a zoological expedition in 1884 on behalf of Büttikofer during which he shot a new species of antelope. The newly discovered species was named Jentink's duiker *(Cephalophus jentinki)* in honor of taxidermist Fredricus Jentink at the museum in Leiden. Like the pygmy hippo, it too is endemic to West Africa and has a relatively limited range of distribution, occurring exclusively in the rainforests of Liberia and the western Côte d'Ivoire. Little is known about Jentink's duiker's way of life. Not even a photograph of the animal in the wild has been published. The banded duiker, another antelope found in primary rainforest, has a similarly limited range. In some forested areas of Liberia the banded duiker is supposedly very common although equally little is known about its biology in the wild. Among the larger mammals of the rainforest, a species of mongoose probably remains the most mysterious. Recently discovered in northeast Liberia in 1958, local hunters probably know most about just where *Liberiictis kuhni* can be found.

Much of today's knowledge of mammals in West Africa can be traced back to zoological expeditions in Liberia during the nineteenth century. Büttikofer's travels are the best known, not the least because he was connected with an important museum. He also recorded his journeys with interesting and informative essays. Büttikofer was not an insensitive hunter: "When a mother monkey is shot and falls to the ground holding her baby she does her best to protect it even in the face of death. Holding her offspring tenderly, she will grimace or meet the cruel hunter's eyes with a pleading look according to the circumstances. I have seen monkeys die with such a pleading or accusing expression on their faces that I have felt like a murderer and have suffered from a bad conscience for some time afterward."

The Diversity and Biomass of Mammals

Although zoologists have been quite thorough in their inventory work on mammals, little is known about the biology of species difficult to observe. Even the geographical distribution of many species in West Africa remains a subject of discussion. Adequate lists of mammal species in specific areas were published for the first time just a few years ago. Over the past years, Harald H. Roth and his students have done a number of studies in Taï National Park in the western Côte d'Ivoire, a part of the species rich forests of the Upper Guinea block [70,71]. Taï National Park covers an area of 3300 square kilometers and lies completely within the rainforest zone. It forms the core of the "West" center of endemism and is probably representative for its original diversity of plant and animal life. Since areas of secondary forest exist in Taï today, a few savannah species of rodents and bats are also found there. Studies have shown a total of 140 mammal species: 14 species of insectivores, 43 species of bats, 11 species of primates, 3 species of pangolins, 41 species of rodents, 11 different carnivores, 15 even-toed ungulates, as well as the African elephant and the tree hyrax (see Annex 2). So whoever believes limited plant diversity necessarily implies a limited diversity of fauna is far from correct: It is true that in the highly diverse rainforests of Panama, a greater number of mammal species has been recorded but that high figure is due to an overproportion-

ally large number of bats. In comparison to Latin American forests, African rainforests are richer in species of large mammals (Tab. 7). 93% of the 150 mammal species known to occur in the Upper Guinea forest block are found in Taï National Park. Nearly one third of those species are restricted to the Guinea forest block and 12 species occur only in Liberia and the western Côte d'Ivoire, in the "West" center of endemism [71]. Over thousands of years, these species have not spread beyond the limits of their ancient range. And it is their limited range which makes them most vulnerable to forest destruction.

Where many different species share a common habitat, one can expect few specimens of each to be present in a given area. The population density of each species is low. Population densities are generally assumed to be low in tropical rainforests and since many of the mammals are small, their total weight (biomass) is also believed to be relatively low. The mammal biomass in Malaysian rainforests has been estimated at 10–15 kg/hectare [72]. In the rainforests on the island of Barro Colorado (Panama), estimates of 17.75 kg/hectare have been made, more than half of which consists of tree-dwelling mammals [72]. But it is not easy to judge the total weight of animals in any given area of rainforest. Reliable estimates of population density have been made for very few species. And since the density of animals in the rainforest is generally underestimated and perhaps even greatly

Table 7
Species of mammals in Taï National Park, Côte d'Ivoire (see also Annex 2) in comparison with much larger rainforest areas in Panama [72] and Malaysia [73].

Rainforest area	Bats	Rodents	Other species	Total
Taï Rainforest	43	41	56	140
Panama	100	48	48	196
Continental Malaysia	86	55	65	206

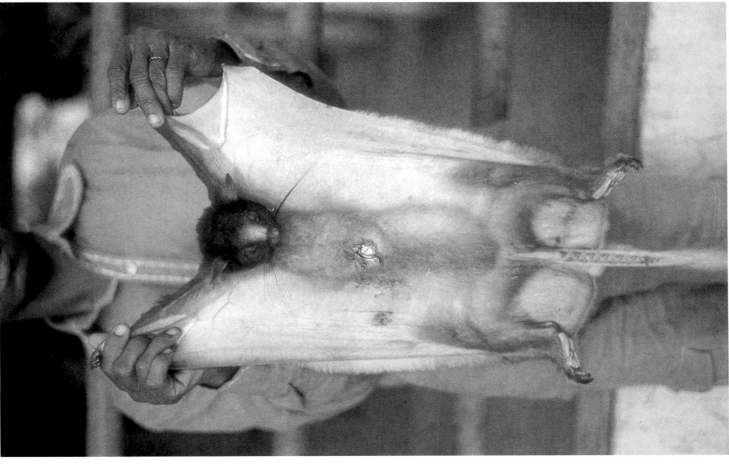

so, estimates of biomass remain random and inexact. It is possible that the biomass in African rainforests is much higher than in Panama. In many West African rainforests, the elephants alone make up 6.5 kg/hectare. A further 3.0 kg/hectare are taken by the trio of tree-dwellers red colobus, black-and-white colobus and diana monkey. But they are only three of the eight to nine species of monkeys and two species of lower primates in undisturbed forests. The higher the proportion of ground-dwelling mammals, animals of greater weight, the higher the biomass. African rainforests are especially rich in ground-dwelling ungulates which is apparent in the catch most often brought in by local hunters – duikers.

This Pel's flying squirrel was killed by a Ghanaian hunter's shotgun. Because of the horny scales at the base of the tail, African flying squirrels are also called "scaly-tailed squirrels". They belong to an otherwise extinct group of rodents. The scales help them to take a firm hold on tree trunks.

Ecological Niches
in Undisturbed and
Disturbed Forest

Ten different species of squirrels and flying squirrels share the habitat on the shore of a small African forest river (see also illustration on previous page).

1 *Fraser's flying squirrel*: Open forest, tree hollows. Eats leaves, fruits and flowers.

2 *Pel's flying squirrel*: Closed forest, tree hollows. Eats the fruit of the oil palm and other fruits. Nocturnal.

3 *Pygmy flying squirrel*: Lives in groups in tree hollows, in closed forest. Probably eats seeds.

4 *Small forest squirrel*: Middle canopy levels, also on the forest floor and in plantations. Eats fruits, nuts and insects.

5 *Red-legged sun squirrel*: Middle and upper levels in closed forest. Eats fruits, oil palm kernels, insects and dead birds.

6 *Palm squirrel*: Ground-dwelling, near rivers and raphia swamps. Eats mainly fruits.

7 *Small green squirrel*: Forest edges, secondary vegetation. Eats the fruit of the oil palm.

8 *Giant forest squirrel*: Canopy-dweller in closed forest. Eats nuts and fruits.

9 *Striped ground squirrel*: Ground-dwelling in clearings and opened areas. Originally a savannah species.

10 *Red-footed squirrel*: Forest edge, secondary vegetation, also ground-dwelling. Eats fruits, including those of the oil palm.

A tropical rainforest seems to offer an inexhaustible food supply, at least for leaf-eaters and especially for those able to climb. But appearances deceive. Leaf-eaters must use as much care in choosing their meal as any hiker does in looking for edible plants in the wild. The green inhabitants of the forest know well how to protect themselves against being eaten. Most woody plants contain large quantities of tannins or other phenolics making them inedible or impossible to digest. Herbs, on the other hand, often contain complex poisonous substances called alkaloids. In a test conducted in two areas of African rainforest, only 10% of the leaves sampled did not contain either tannins or alkaloids. Many plants growing on soils poor in nutrients and on sandy soil seem to contain especially high quantities of tannins and phenolics since these substances do a good job of keeping plant-eaters at a distance [74]. A high fiber content can also serve as a good protec-

tion [75]. It seems to be especially important for plants on poor soil to protect themselves since the lack of nutrients (nitrogen, phosphorus and minerals) makes it difficult to replace lost leaves.

A population of black colobus (*Colobus satanas*) is found in the Douala-Edea Rainforest Reserve in Cameroon, where the soil is very sandy and low in nutrients. In contrast to other species of colobus, black colobus do not feed mainly on the fully grown leaves of trees common to the Douala-Edea forest, but on the leaves of rare deciduous trees and of climbers. Seeds make up 53% of their diet. Thus, in spite of the rainforest's rich vegetation, the food supply can be low which is probably why the population density of the black colobus in the Douala-Edea Reserve is only one tenth that of colobus species in other regions [76]. And competition between species further complicates the situation.

Avoiding Competition

Standing on the floor of the forest, one can barely imagine the incredible variation and diversity of the surrounding habitat for the various niches are not easy to perceive. From the hollows beneath the termite nests and between the roots of large trees to the forked branches of the giants towering above all else, the spacial division of the forest is barely visible. We have seen how epiphytes live on the branches of trees. It would surely be interesting to learn more of how bats, flying squirrels, pangolins and other animals make use of hollow trees, if only they could be better observed.

The tropical rainforest is unique in its vertical structure which is far more developed than in temperate forests. In comparison to the upper levels, the forest floor is rather monotonous. The fact that such a high proportion of animal species are tree-dwelling mirrors the diverse vertical structure of the rainforest from the floor to the uppermost tree tops. This in turn contributes to the especially high diversity of species.

Apart from the spacial division of the forest, there are a number of other factors which allow different animal species to coexist: separate diets or different periods of activity within the same habitat for example. The term "ecological niche" therefore does not apply only to spacial division. It encompasses all the factors concerning when and how a habitat is shared by various species and includes spacial distribution, utilization, and the season or time of day of activity. Neither is the distribution of ecological niches among species, the niche separation, only a question of preference. In a habitat occupied by many different species, no single species occupies space and resources as it would in an "empty" forest. Competition influences which niches are occupied by the species present. And the more similar the demands of different species, the tougher the competition for niches will be. Dominant species will push others into less favorable niches and this can influence their population density. In extreme cases, one species may push another entirely out of a certain habitat. Competition between species also accounts for the fact that the same species may be found occupying different ecological niches in different areas. Where the competition is low, however, a species will take advantage of the situation and expand resulting in its holding a monopoly on the "market". The situation may well be compared to that of a free market economy: When a beer brewery holds a monopoly, it will do as it pleases. But when the market needs to be shared by one hundred different brands of beer, each brewery must be prepared to enter into agreements and make compromises, to join the cartel and specialize itself to fit into a specific market "niche" in order for the system to function properly.

A keen observer in the rainforest can at best get a small idea of the "cartel". It would be impossible to investigate all the "agreements" and "compromises" between species in the rainforest. Lacking such specific knowledge, we can only wonder at the fact that no two species have identical demands. It is almost as though each were assigned its niche together with certain rules of behavior. It is especially important for closely related species to occupy separate niches. Their similar body structure and diets could otherwise lead to dangerous competition and a serious threat to the existence of one or the other. In West African rainforests, there are many groups of closely related species which survive only by means of an admirable system of niche separation.

Seven different species of squirrels occur in the moist semi-deciduous forests of western Ghana [66]. Little is known about the ecology and behavior of individual species but since squirrels are mainly active during the day, they are relatively easy to observe even in closed for-

tates an even better system of niche separation within the forest.

Different Methods of Spacial Utilization

Niche separation, which allows different species to occupy the same habitat, does not do away with the danger of competition. Within the same species, competition for edible and palatable food items can also lead to problems. For those species living in groups, this kind of intraspecific competition poses a threat to the individual as well as to the community. If a species is not able to deal with the situation, it will overuse its food supply as surprising as that may seem for a rainforest. The consequence would be a scarce food supply resulting in a reduction of the number of individuals or, even worse, a collapse of the community, which would not be in the interest of the species. How does a species solve this conflict: Keeping the population density up while simultaneously preserving the food supply? Two strategies may be recognized to avoid competition between members of the same species [72]:

1. Solitary animals have limited home ranges or territories which they defend against other members of their species. This results in an even distribution of individuals. Many ground-dwelling species follow this pattern. In West Africa, duikers for example live either solitary or in small family groups.

2. Gregarious species living in groups move from place to place over larger areas according to the food supply. They share group home ranges which may overlap with the home ranges of other groups. Many tree-dwelling species of monkey follow this pattern of spacial distribution.

The more animals present in a group, the more mobile it has to be. It needs to move from one food supply to the next, always finding enough

est, once the problem of species identification has been solved. One soon notices that certain species are usually seen on the ground, others only in the branches and still others only outside of closed forest areas (Fig. p.112) Many of these species have a preference for the fruits and nuts of the oil palm, a tree native to West Africa. Otherwise, however, each species appears to have a somewhat different diet [77]. The spacial distribution and division of food resources is similar among monkey species in West African rainforests (Fig. p.114). But monkeys live together in larger clans, a situation which necessi-

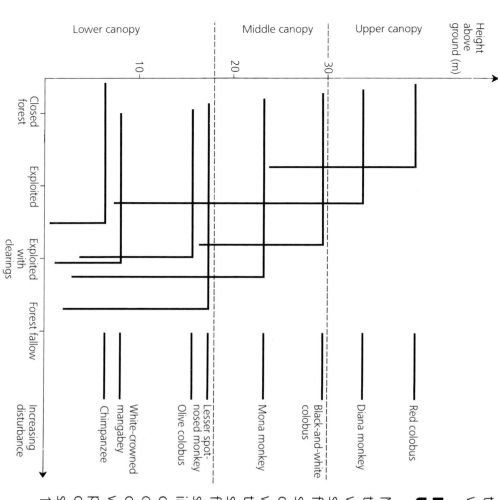

Spacial utilization by eight higher primate species of moist semi-deciduous rainforest in western Ghana. Two lower primates, the potto and the dwarf galago, can also be found in this area of rainforest. They are not included in the diagram.

In time, the primate species of the upper canopy levels become habituated to human presence on the forest floor and are relatively easy to observe.

Studies of Sympatric Monkey Species

A guttural "rurr-rurr-rurr ..." awakens the visitor to Bia National Park's research station in western Ghana at just about 6 a.m., piercing the early morning stillness as though someone had started up an old diesel engine. One can scarcely determine from where the peculiar sound originates. It echos through the twilight of the forest to rebound weakly from all directions. After 10–15 seconds, the eerie noise stops as suddenly as it began. Day has begun for the black-and-white colobus. The male leaders use calls to

to cover the current food requirement of the entire group. The best example of an abundant food supply in the tropical forest are trees in blossom or bearing fruit. A giant tree covered with thousands of fruits does not attract the members of one species alone: It will temporarily become a meeting place for all kinds of fruit-eaters. All sorts of animals will amply feed day and night in the branches or search out fallen fruit on the ground below. This massive congregation of fruit-eaters is often vital for the tree's own survival. Seeds are ingested together with the fruit and "sown" elsewhere with the animals' dung.

A subspecies of black-and-white colobus, *Colobus po-lykomos vellerosus*, occurs in Bia National Park (cf. also Fig. p.40). Colobus have stunted thumbs as can be observed on this poached specimen (left).

mark their clan's position at the start of the day. The one group's signal causes the next to mark its position and the rolling sound quickly moves through the forest like an order of the day. Several groups of this species were observed over periods of months and years during which group development, movements within the group home range and diet were recorded. The studies showed that a forest full of monkeys actually reflects a very highly organized system: The movement patterns and feeding habits of monkeys are far from random. In Bia National Park near the western border of Ghana in the

area of transition from moist evergreen to moist semi-deciduous forest, groups of other monkey species were also closely studied during 1974–78. Eight higher species of monkey are sympatric here, meaning they occur in one and the same area of forest. The Ghanaian Department for Game and Wildlife under the direction of Emmanuel O. A. Asibey took great efforts to place representative areas of rainforest under permanent protection during those years. The areas were studied as best possible with the limited funds available. The rapid rate at which Ghanaian forests were being opened up for

Red colobus move with perilous jumps through the canopy, one after the other and are thus easy prey for local hunters.

timber exploitation had left few larger forested areas undisturbed and those areas remaining came under increased pressure from the timber industry. The author of this book was employed as a national park director in a project supported by WWF and the International Union for Conservation of Nature and Natural Resources (IUCN) for the management of the protected areas in the Bia region. He was later also involved in a conservation program for the Ankasa area in the southwestern part of the country. The primate studies, all conducted by American researchers in the center of Bia National Park, were also a

part of this rainforest conservation program. The American scientists worked with the continual assistance of up to 18 trained local aides and were so able to record nearly all the activities of three groups of black-and-white colobus, two groups of red colobus and one group of diana monkeys within the same study area. After a difficult initial phase, these upper canopy monkey species became habituated to their human observers who followed them from below every step of the way through the forest. The monkey species of the lower canopies are more difficult to habituate to observation. Horizontal visibility

is much less and the monkeys flee through the undergrowth just as soon as one catches a glimpse of them. Probably nowhere else in West Africa's rainforests has the ecology and sociology of sympatric monkey species been as closely studied as in the Bia area of western Ghana. In the Kibale forest of Uganda at 1500 meters above sea level, however, intensive studies of sympatric monkey species had been in progress since 1970. Carried out in the area of transition to mist forest, the studies focused mainly on red colobus and guerezas (Colobus guereza) [78,79]. In the Kibale forest, the guereza fills a niche similar to that of the black-and-white colobus in the Bia area of Ghana. In the 5.4 square kilometer study area of Bia National Park, observers guided themselves by means of a numbered grid of transects. The grid of narrow trails cut through the undergrowth of the forest with cutlasses, measured a total length of 58 kilometers. The monkey groups studied were all observed for hundreds and in some cases thousands of hours which resulted in an immense amount of data. One of the groups of black-and-white colobus was observed for a total of 3500 hours. It is perhaps due to the sheer quantity of data material that no comprehensive publication has appeared to date. Only some data concerning group interaction among black-and-white colobus have been published [80]. As unfortunate as it may be, the Ghanaian Department for Game and Wildlife, which supported the studies, was not merely interested in their scientific aspect but also in estimating the animals' population density and the economic value of intact forest for the local human population. For in West Africa, monkeys are of scientific interest only to a small foreign minority. The local majority is more interested in the animals' gastronomical aspects. As long as these primate species are not over-hunted and their numbers do not drop below a certain minimal level, they serve as an important and sustainable food source for the local population. The precise information collected on the groups of the three crown canopy species in the study area and on their home ranges at least allowed important comparisons to be made with areas of disturbed forest surrounding Bia National Park [81].

Spacial Organization of Primates in the Crown Canopy

It is not by chance that of the various monkey species in the rainforest, scientists choose to study the inhabitants of the upper layers: At least in West African rainforests, it is easier to get a view of the upper levels of the forest through openings in the canopies than of the lower levels. The lower reaches of the forest may be closer to an observer on the ground, but they are well hidden by the dense growth of saplings, intertwined vines and other climbers. Since the three crown-dwelling species of monkey never set foot on the ground, they must also fulfill their water requirement by means of their food intake. After a rain shower, of course, small pools of drinking water collect in tree hollows and in the leaf rosettes of epiphytic plants.

Red colobus monkeys are the most talented inhabitants of the uppermost canopy in West African rainforests. They move in large groups through the upper levels of the forest, often jumping amazing distances from tree to tree. An entire group will follow its leader single file springing from the same branch to the next tree top. An observer standing in the right place beneath two towering trees can easily count the members of a group as they jump. The larger of the two groups studied in Bia National Park consisted of 58 members. A group of such size creates a caravan of monkeys stretching through several trees accompanied all the while by a chorus of grunts and barks. Their calls turn into panicky screeches should a crowned hawk-eagle (Stephanoaetus coronatus) be sighted

⌄ The mona monkey occurs in the lower levels of the forest, where it feeds almost exclusively on fruit. It adapts to disturbances in the forest and may even venture in groups into plantations – not to the pleasure of farmers.

gliding just above the tree tops. With lightening agility, the monkeys flee to the lower levels of the forest, sliding down trunks or jumping di-rectly from the branches into the forest below. It is not unusual for red colobus to freefall 10 to 20 meters in order to escape their deadly enemy in spite of the risk of crashing through to the ground. Skeletons of upper canopy monkey spe-cies often show healed bones apparently once broken by such hard landings. The crowned hawk-eagle is just about the only predator dan-gerous to the red colobus. The feces of leopards in Taï National Park, however, have also been known to contain red colobus hairs beside the traces of many other prey [82]. The population density of leopards in the rainforest is low but they apparently make their way into the highest branches in search of prey. Human hunters have little problem in shooting red colobus as they are easily spotted due to their frequent vocaliza-tion. When the monkeys flee in single file, they can be shot one after the other. In areas of heavy hunting, the red colobus is thus the first species to be wiped out.

Black-and-white colobus lead less conspicuous lives than their relatives in the "Red Brigade" of the tree tops. They live in small to middle-sized groups of up to 25 individuals and tend to keep to the middle levels of the forest. Often hidden within thickly entwined lianas close to the trunks of trees, they are not easily spied by human observers. This way of life, however, may be caused by the presence of red colobus. In regions where the latter does not occur, black-and-white colobus supposedly make their way into the uppermost canopy. They have a strong preference for leaves and are able to survive in secondary forest and even in small, isolated for-est plots in farmed areas. Monkeys like the black-and-whites relying on abundant and evenly distributed food resources live in closely knit groups and do not need very large home ranges [79].

The third West African primate species inhabit-ing the tree tops is the diana monkey. An agile and apparently adaptable species, diana mon-keys have quite a varied diet; they also feed on smaller animals, mainly arthropods. This species with its attractive coloring and long, pointed beard lives in groups of 10 to 20 individuals. It tends to spend its time at the middle levels of the forest although it may on occasion climb higher or descend to the lowest levels. Diana monkey groups not habituated to human pres-ence are very difficult to observe. They move through the forest quickly and occupy large home ranges.

Few studies have been made on group interac-tion and the feeding habits of the mona monkey, the lesser spot-nosed monkey and the olive co-lobus. Little is known about these species although they also inhabit areas of secondary vegetation and are therefore more widely dis-tributed today. The white-crowned mangabey, a subspecies of mangabey which occurs only in the eastern part of the Côte d'Ivoire and in Ghana, and the chimpanzee are almost impossi-ble to observe since they flee through the undergrowth across the forest floor leaving even a keen observer behind without a clue.

The eight primate species present in the study area within Bia National Park divide and share a single habitat in a complex manner (Fig. p.114). In order to truly understand how competition between and within the various species is avoided, close studies of all eight species would be necessary. Where and when individual spe-cies are present would have to be related to their feeding habits. These in turn would have to be compared with the food availability, which may vary geographically and seasonally. And fi-nally, all these variables concerning eight differ-ent species would have to be interrelated and considered as a whole. Considering that mon-keys feed on dozens to more than a hundred different plants, an analysis of their feeding ecol-

A farmer has killed a lesser spot-nosed monkey in a thicket at the edge of a slash-and-burn clearing. This monkey species often forms mixed groups with mona monkeys.

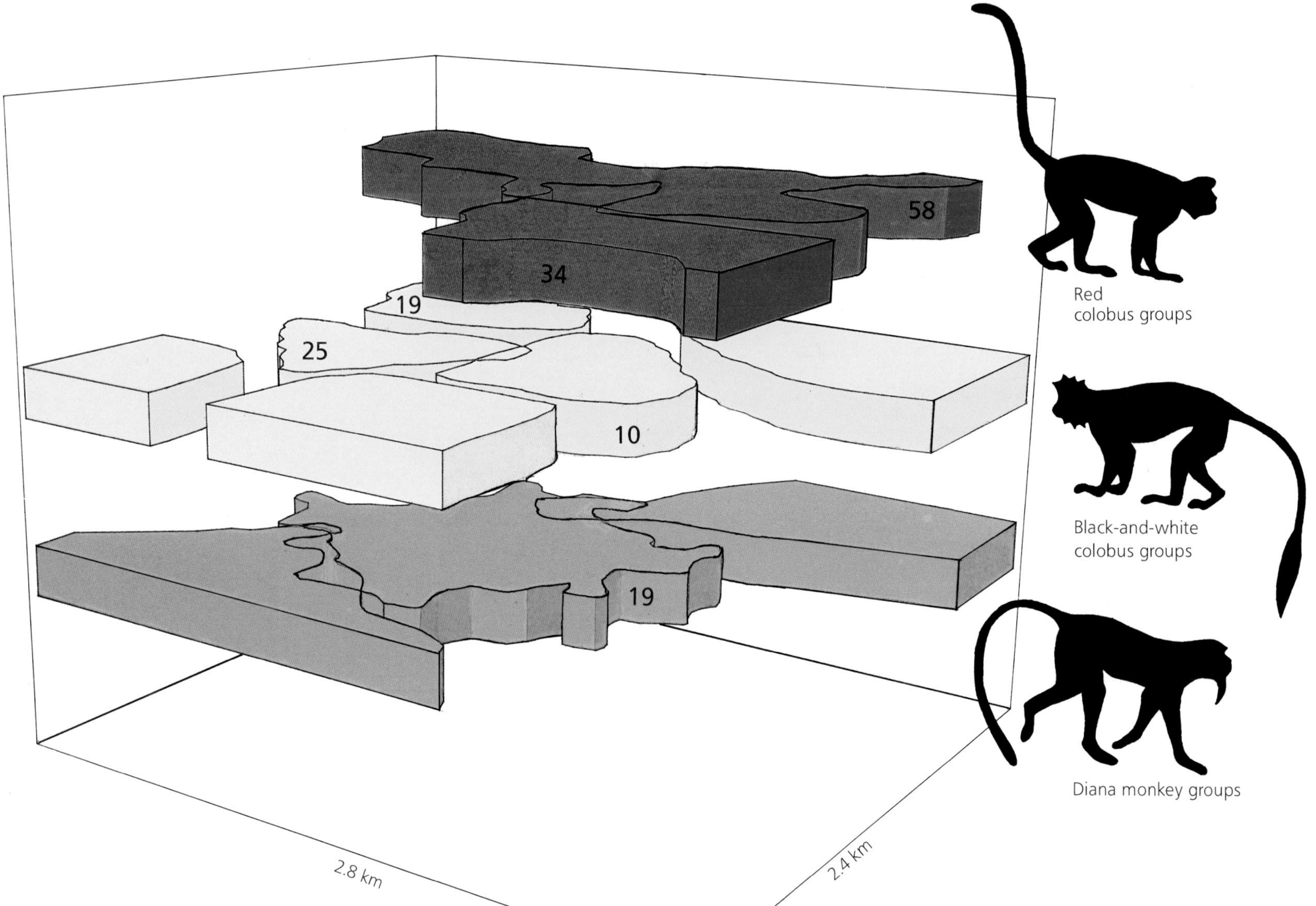

Home ranges of groups of canopy-dwelling monkey species in the study area of Bia National Park. The numbers refer to the members in each group observed. The diagram has been somewhat simplified. In reality, the home ranges of the three species are not as clearly separated vertically. (cf. also Fig. p.114).

2.8 km

2.4 km

19

25

19

10

34

58

Diana monkey groups

Black-and-white colobus groups

Red colobus groups

Chimpanzee food – the "African walnut" (*Coula edulis*).

Spacial Utilization by Three Upper Canopy Monkey Species in the Study Area of Bia National Park

Red colobus mainly feed on leaves but also on the fruits and seeds of more than 50 middle-sized to tall trees. In addition, they feed on the leaves, fruits and flowers of at least 15 climbers [83]. Over the course of one year, the larger of the two groups studied (58 individuals) occupied a home range area of 1.9 square kilometers. The home range of the smaller group (34 individuals) measured 1.4 square kilometers. This resulted in a population density within the study area of 24.8 red colobus per square kilometer.

Black-and-white colobus have a strong preference for leaves: They feed on 120 different species of plant, of which 91 are trees [84]. They live in small to middle-sized groups of 10–25 individuals and the home range areas measure approximately 0.7 square kilometers. The population density within the study area was calculated at 21.7 black-and-white colobus per square kilometer.

Diana monkeys take advantage of many food sources: In Bia National Park their diet consists of the leaves, fruits and oil-rich seeds belonging to 153 plant species, but neither do they shy eating animals, especially arthropods [84]. The 19 individuals in the group studied occupied a home range of 1.5 square kilometers. The population density within the study area was calculated at 12.8 diana monkeys per square kilometer.

ogy including so many variables would be an impossible task without a powerful computer. And any analysis would be further complicated by the fact that monkeys often feed only on specific parts of a plant: young or old leaves, stems, fruits, flowers or nectar. But only a detailed study of all these aspects would allow us to really claim to understand the niche separation of the primate species in this area of rainforest. We will have to satisfy ourselves with less for the moment: a look at the spacial utilization of the upper canopy by its three monkey species inhabitants (Fig. p.122). Within the study area of Bia National Park in 1978, approximately 60 individuals of these three species were present per square kilometer. In the Kibale forest of Uganda, the upper canopy species live in groups of simi-

lar size, but require smaller home ranges than in Bia National Park. The number of red colobus per square kilometer in Kibale is thus equal to the total for all three upper canopy species in Bia [79].

Nut-cracking Chimpanzees

Two Swiss researchers, Christophe and Hedwige Boesch, have made some extraordinary discoveries concerning chimpanzees in West African rainforests. They successfully managed to habituate wild chimpanzees to their presence and have been studying their behavior since 1978 in the closed forest of Taï National Park in the southwestern Côte d'Ivoire. The Boesches have paid particular attention to the chimpanzees' use of tools.

The local human population must have long known that chimpanzees use simple tools in their search for food. In Büttikofer's reports on his travels last century he wrote: "The baboon – as the chimpanzee is called in all of Liberia – is considered to be more highly developed than other animals." Chimpanzees were not hunted and eaten as were other primate species: "The people say that the baboon is too much like man" [69]. Even today, some forest-dwelling people do not hunt chimpanzees. But the scientific community did not learn of the chimpanzees' ability to use tools, a characteristic which had otherwise been solely attributed to man, until 1964 when Jane Goodall made such observations in Tanzania. She described how chimpanzees used small sticks and bits of straw to fish termites out of their nests.

In 1971, two scientists researched chimpanzee behavior in the southwestern Côte d'Ivoire near the villages of Troy and Sakré not far from the Liberian border. They found places where chimps had used tools to crack nuts [85]. The Boesches systematically studied this phenomenon [86]. In an area of 13 square kilometers of

Chimpanzees in Taï National
Park use a wooden hammer
on this anvil to crack open
the very hard nuts of *Panda
oleosa*. The hammer shows
visible traces of wear.

lowland rainforest in Taï National Park, they found more than 1400 anvils around which lay the shells of cracked nuts. These are places where chimpanzees use wooden clubs and stones to crack the nuts of five different kinds of pitted fruits. Among them are some of the hardest nuts known in the forest. They are embedded in the fruit pulp which soon rots away and contain one or several kernels within the hard shell. Experiments have shown that a stone weighing 10 kilograms has to be dropped from a height of 120 centimeters in order to crack the large nut of the middle-sized tree *Panda oleosa*. The chimpanzees achieve this task with smaller stones and sometimes only with wooden clubs or natural hammers to pound the nuts. They place the pits or nuts in small indentations on roots protruding above ground or on rock surfaces so that the hammer repeatedly strikes the right spot on the nut in order to crack it open and not just smash it to bits. In time, the indentations on the anvils become deeper and the tools themselves show similar depressions. After first finding the necessary tools, the chimpanzees often have to carry them over long distances to where they can be used.

The chimpanzees often open the nuts of *Coula edulis* (Fig. p.123) while sitting in the branches of the tree. Although these nuts are smaller and not as hard, the chimps still need some sort of tool to open them. For this reason they carry their hammers with them when searching for nuts in the trees. Coula nuts have a pleasant taste and are often sold at local markets under the name "African walnut". Anvils for other kinds of nuts (*Parinari excelsa, Sacoglottis gabonensis* and *Detarium senegalense*) were noticeably less common although Parinari and Sacoglottis trees are far from rare. Chimpanzees appear to prefer the larger nuts thus adding proteins and fats to their otherwise varied diet of leaves.

An interesting aspect of these observations is not only the chimpanzees' use of tools but also

the fact that the apes seem to be able to plan their activities and combine certain acts to achieve their goal of getting at food: They gather the nuts in their mouths and hands, carry them to an anvil and then find a suitable hammer to crack them. Stones are not often found in these rainforest areas. In order to crack the hard Panda nuts, the chimps must borrow stone hammers from other anvils and carry them over long distances before they can open and eat the nuts they have collected. The chimpanzee's ability to use its environment, the rainforest, in this manner is apparently an acquired trait passed on from generation to generation.

The Effect of Timber Exploitation on Rainforest Fauna

Certainly, no other habitat on earth is as rich in animal species as an undisturbed rainforest. But large undisturbed forest areas have become rare in West Africa today, where timber exploitation and slash and burn clearing are so common. With the opening of the forest by timber companies and the resulting invasion of farmers, increased hunting pressure led to the local disappearance of many species. In large areas of secondary forest and fallow land, the fauna today is seriously impoverished.

Even selective timber exploitation leads to the decimation of species diversity in many faunal groups. A great number of mammals cannot survive the opening of the canopy: The bay duiker, Jentink's duiker, Ogilby's duiker, the yellow-backed duiker and the banded duiker are all more or less dependent on closed primary forest. The pygmy hippopotamus, the giant forest hog, the water chevrotain and a number of insectivores, rodents and bats are very sensitive to changes or disturbances in forest structure. 48 of the mammal species occurring in the Taï region of the Côte d'Ivoire are almost exclusively dependent on closed forest [71].

Thanks to its similarity to man, the chimpanzee is often spared by hunters. The fate of these apes is much more entwined with that of the forest. Although chimpanzees also occur in some parts of Africa's tree savannah, they are mainly forest-dwellers.

Primates are especially vulnerable to disturbances. Not only do they suffer from a changed forest structure, but heavier human settlement in opened areas leads to increased hunting pressure which, monkeys being easy prey, can quickly lead to a species being wiped out. Unfortunately, it is often difficult to prove the negative effects of exploitation and increased hunting since there is usually no way of comparing the situation to that in an undisturbed area.

Lower Monkey Diversity in Secondary Forest

The years of research on the monkey species of the upper canopy in Bia National Park allowed comparisons to be made between monkey populations in the study area known to be undisturbed and populations in adjacent areas disturbed to varying extents [81]. The entire region is part of one lowland rainforest which had stretched uninterrupted over a large area until timber exploitation began in 1973. A trained observer recorded the presence of monkey groups in all areas of the forest during 13 months between 1977 and 1978, covering 978 kilometers on foot. Together with an aide, he regularly walked the transect paths cut through the undergrowth of the forest for purposes of observation. Within the study area of Bia National Park, the lines of observation had already been cut and marked, one only needed to randomly choose a similar stretch for comparison.

The probability of sighting monkeys varies with the time of day. Therefore, when comparing two areas, it is very important that observations be made during similar hours and not here in the morning and there at noon. A meaningful comparison of two different areas can in any case only be made using strictly standardized methods of observation. And even then, leeway must be allowed for uncertainties when observing in a rainforest. A trained eye is capable of observing monkey groups in the upper canopy 50 to 80 meters right and left of a transect line without error. Monkeys groups at the lower levels can at best be detected up to a distance of approximately 30 meters and it is difficult to judge the size of low level groups. Mona monkeys and lesser spot-nosed monkeys moreover often form mixed groups. An observer must note them as being one group as if only one species had been sighted. Finally, chimpanzees are very secretive and difficult to observe despite their size. At the approach of a human observer, the shy apes will flee and disappear quickly and quietly through the undergrowth. The actual number or population density of a species cannot be judged merely on the basis of such observations. The frequency of sightings in different areas, however, may be compared.

The Bia study delivered clear results and provided statistically significant differences (Tab. 8): Inside Bia National Park, which is undisturbed and protected against poachers, a keen observer is likely to sight an average of one monkey group for each kilometer covered. The neighboring Sukusuku Forest Reserve has been subjected to timber exploitation and is partially farmed. In this area, an observer is likely to sight one monkey group over a distance of 2.2 kilometers. The red colobus is no longer present and the groups of black-and-white colobus and diana monkeys are noticeably smaller.

Since timber exploitation usually aims for the tall and emergent trees, one can expect the monkey species of the upper canopy to be the most disturbed. When in 1977, part of the Bia Forest Reserve was subjected to selective timber exploitation, the entire populations of red colobus and black-and-white colobus fled to neighboring forest areas. This led to temporarily increased monkey group densities in those areas. Once the exploitation had ended, black-and-white colobus returned to the young secondary forest. So although higher degrees of disturbance lead to fewer groups of black-and-white colobus being sighted, this particular species is surprisingly resilient in the face of timber exploitation and even hunting pressure. Its strength perhaps results from its sedentary way of life and the fact that it hides well in tangles of vines and other climbers.

Red colobus have more difficulty with opened-canopy forest. For these inhabitants of the uppermost levels, even a selective extraction of timber greatly alters their habitat. Not only in Bia, but in the Gola Forest Reserves of Sierra Leone and in the Taï forest area of the Côte d'Ivoire as well, it has been seen that red colobus vanish once timber exploitation sets in [87,71]. Since timber extraction is widespread throughout tropical forests in West Africa, this species of monkey must be considered severely endangered. The transport routes and feeder roads for timber extraction also allow hunters easier access to the forest adding increased hunting pressure to the problem of habitat alteration. Diana monkeys are also sensitive to the combination of disturbances in the forest. In Ghana's

Bia region and in the Gola Forest Reserves of Sierra Leone, they do occur in secondary forest but soon disappear once hunting pressure increases. Today, the diana monkey is also an endangered species.

Ground-dwelling monkey species and those of the lower levels are also sighted somewhat less often in secondary forest than in undisturbed primary forest, but just how disturbances affect them is less clear. Mona monkeys and lesser spot-nosed monkeys prefer thick undergrowth in small gaps of the forest. Smaller groups can survive in the thickets of forest fallow areas in and around plantations when they are not killed off by hunters. There are no indications, however, that population densities in disturbed areas are greater than in closed forest. Finally, the olive colobus lives hidden away in thickets and is always difficult to observe. It supposedly prefers swampy forest [71,87]. Although it has been considered to be endangered, it may be better off than the top canopy species as it clearly survives in secondary forest as well.

Table 8
Number of monkey groups observed per 100 km walking distance in five zones of the Bia area in Ghana [81]

Monkey groups	Bia National Park	Bia Game Reserve North	Bia Game Reserve South	Bia Game Reserve West	Sukusuku Forest Reserve
Observation zone	Undisturbed	Undisturbed	Partially exploited	Undisturbed plantations nearby	Exploited with plantations
	Satisfactory wildlife protection	Moderate wildlife protection	Moderate wildlife protection	Moderate wildlife protection	No wildlife protection
Red colobus	9.8	6.1	0.5	–	–
Black-and-white colobus	28.5	16.6	16.4	15.5	9.8
Olive colobus	5.7	11.0	8.4	0.9	0.5
Diana monkey	17.9	18.3	12.5	11.5	7.2
Mona monkey and/or Lesser spot-nosed monkey	39.1	39.1	32.7	31.9	25.9
White-crowned mangabey	2.4	3.4	1.5	0.9	1.6

Forest mammals which commute between forest and secondary vegetation and enter plantations: red-footed squirrel, brush-tailed porcupine, Emin's giant rat, lesser spot-nosed monkey, mona monkey, forest buffalo, forest elephant.

Savannah species which follow human settlement deep into the rainforest: striped ground squirrel, Gambian giant rat, large cane rat, Egyptian mongoose, large-spotted genet, various mouse species.

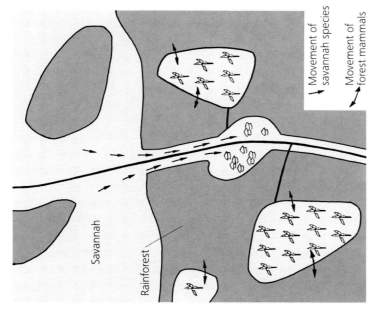

Savannah

Rainforest

→ Movement of savannah species

↔ Movement of forest mammals

Commercial Timber is Monkey Food

Timber extraction not only alters the structure of the forest, it influences the tree species spectrum by removing economically valuable species and allowing others to remain. Since practically all commercial timber species are among the tallest trees with trunks of large diameter, the tree species spectrum in the upper canopy is the most hard hit by the extraction procedure. The upper canopy monkey species are thus also the most affected in view of their diet. Clearly, it is not by chance that the diets of red colobus, black-and-white colobus and diana monkeys consist to a disproportionately large part of commercial timber species. In the Bia area of Ghana, 43% of the species found in the diet of red colobus were economically valuable timber species [83]. The proportion of commercial timber species in the diet of black-and-white colobus was found to be over 25% and included all eight species listed under the Ghanaian prime timber class. Among them are the valuable species of mahogany (*Khaya* spp.), iroko, sapele, utile and makore. The commercial species utile and obeche are particularly important for the black-and-white colobus. Diana monkeys also feed largely on tree species; 20% of those are economically valuable species. Iroko and antiaris are especially important for their diet [84].

So West African timber is not only of use to man. One can assume that other creatures depend more vitally on certain timber species than do humans. Nevertheless, it is probably not the change in the tree species composition which causes the monkeys' worst problems; what really hurts is the change in forest structure. The breaking up of the canopy through timber extraction, followed by more intensive hunting practices and slash-and-burn clearing on the whole drastically affect the monkeys' living conditions.

The Fauna of Secondary Vegetation

Opening up West African rainforests for timber extraction, slash-and-burn farming, increasing areas of fallow land and the ultimate total destruction of the forest result in a drastic loss of species diversity in most of the former forest area. But as is the case everywhere, there are creatures in the rainforests as well which profit from the misfortune of others. Degraded forest provides habitat for those animals which prefer gaps to closed forest vegetation. Degraded forest also makes room for animals dependent on open forest and savannah (Fig. p.129) They often follow human settlement.

When old age causes a giant tree to crash to the ground in a closed rainforest, other smaller trees are pulled down with it to rot on the forest floor. For the first time in centuries, sunlight shines through the gap and actually reaches the

The forest buffalo, the forest form of the African buffalo, is very difficult to observe. It has profited from today's increased area of secondary vegetation, for example in the Côte d'Ivoire. Its area of distribution has increased.

ground below. But not for long. Creepers and climbers soon spring up from between the broken branches. The seeds of climbers and other herbaceous plants long dormant in the shade on the ground or those recently blown into the gap by a breeze have been waiting for just this chance. For a bit of sunlight to fall on the soil where they lie. In no time at all, the fallen trunk and branches are overgrown and soon enough, the first tiny umbrella trees (*Musanga cecropioides*) have germinated. In a few years time, they will cast their shadow over the gap. The naked trunks of trees at the edge of the gap are soon overgrown anew. Creepers and climbers find their way up the trunks and before long a thick green wall of vegetation surrounds the opening in the forest. This is the type of vegetation commonly described as "impenetrable jungle" in many travel accounts. Along the shores of rivers wide enough to allow sunlight through to the ground, similar tangles of leafy vines present themselves to water travellers. But of course, in this case it would not be correct to speak of "secondary vegetation" since it does not result from human disturbance. "Sunlight vegetation" would be a better term for the growth in natural gaps in the canopy, despite the fact that it is quite comparable with the secondary vegetation which springs up in areas cleared by man.

A wide range of animal species regularly visit gaps in the forests and the shores of rivers thickly overgrown with tangled vines, creepers and other herbaceous plants. Birds, which normally keep to the upper canopy, find sufficient sunlight here, flowers, fruits and seeds. In turn, they deposit the seeds of trees with their dung in the clearing and so contribute to the regeneration of the forest. Various squirrel species, the brush-tailed porcupine, the giant rat, monkey species found in the undergrowth and the largest of the forest antelopes, the bongo, also take advantage of the young leaves low in tannins and of the fruits available. All these species belong to those which regularly cross from closed forest into pioneer vegetation, whether it be in a natural gap or in man-made clearings. Should there be a plantation nearby, all the better: Secondary vegetation grows along the edges and the plots offer brush-tailed porcupines and giant rats cassava roots for the digging; monkeys are more interested in the cores of the banana stems and, of course, in the bananas themselves. For an African farmer, there is no worse menace than the forest elephant. It enjoys rounding off its diet with regular visits to secondary forest vegetation or, if possible, to plantations. They leave the plots in a sorry state – yams, coco-yams and bananas are usually uprooted, kicked over and severely damaged. Indeed, small areas of secondary vegetation and plantations enrich the habitat of many animals in the rainforest. For most, however, secondary vegetation is not more than a desirable addition. They remain dependent on closed forest for the most part, whether it be undisturbed primary forest or older, regenerated secondary forest.

The Inhabitants of "Farmbush"

The fast growing umbrella tree (*Musanga cecropioides*) soon fills in gaps and cleared areas; it tends to dominate the edges of timber tracks and access roads. This pioneer can reach a height of ten meters after only two to three years. Its umbrella-shaped leaves cast the first shadows in which the seedlings of true closed forest species grow. Today, the umbrella tree is the most widely found species in all of West Africa. It lines the edges of forest patches, fills in freshly cleared areas quickly and protects the soil against erosion. The umbrella tree is at once a sign of warning and of hope.

Were one to allow umbrella trees to grow as they would, a secondary forest of varied vegetation would eventually develop over the years and decades. But land is usually cleared for a specific purpose: crops. The umbrella tree is of no use to a farmer; either he does not allow it to grow in the first place, or he cuts it down in order to free overgrown timber loading areas and transport routes for farming. Once pioneer vegetation has been cut, the chances of forest regeneration are lost, or at least greatly reduced. So-called "farmbush" develops after only a few years of agricultural use; crops can no longer be cultivated on the depleted soil. The only vegetation able to survive in these fallow areas consists of scrub and grasses. Sloped areas become eroded and may even remain barren. Today, most of West Africa's former forest area is covered by farmbush of this nature.

Due to the resulting increase in grassy areas – a kind of moist savannah – savannah animals have found their way into the rainforest. In addition to some bat species, rodents especially profit from the change in vegetation and appear in the wake of forest destruction. Not that they are inferior to those animals which left. Hunters, at least, know the value of the meat of some. The cane rat, also known as the grasscutter, is appreciated throughout West Africa for its tender white meat. Hunted and trapped in coarse, grassy areas, it can grow to a length of 50 centimeters and may weigh up to seven kilograms. In turn, certain predators which normally live in the savannah are attracted to the rainforest area by the rodent population. The Egyptian mongoose (*Herpestes ichneumon*) or the large-spotted genet (*Genetta tigrina*) for example.

The Forest Buffalo Moves In

West African farmers fear meeting up with one newcomer to the forest – the "bush cow", more scientifically known as a subspecies of the African buffalo (*Syncerus caffer nanus*) adapted to forest habitat. The forest buffalo is smaller than its savannah relative, its horns are not as large and it is reddish in color instead of black. But it possesses the same unpredictable temperament and occasionally causes injuries or worse among the local human population. In transitional areas between forest and savannah, there are intermediate forms of these two African buffalo subspecies.

In spite of its adaptation to forest habitat, the forest buffalo has retained its preference for grass and seldom ventures into extensive primary forest. It keeps to secondary forest where gaps offer grass for grazing and frequents forest bordering on transitional areas to savannah territory. In the Côte d'Ivoire, the distribution range of the forest buffalo has increased since the beginning of the century due to forest loss. It now extends throughout most of the rainforest zone [88]. Today, secondary vegetation is the rule in this country where only few areas of closed forest have survived the rapidly expanding timber exploitation and resulting agricultural activities. The buffalo now occurs in 82% of the forest zone and in 99% of the zone of transition to savannah. The population is low, but more or less continuous. In actual savannah territory, however, the species has lost some of its range due to increased agriculture. It is found in only 66% of the savannah zone. A total population of 50,000 individuals belonging to both species is estimated to be present in the Côte d'Ivoire today. About half the population is present within the rainforest zone [88].

Clash with a forest buffalo last century. The white hunter lives up to his stereotype of being a lifesaver.

The Coevolution of Plants and Animals

Color, form, fragrance and the nectar of flowers are meant to attract pollinating animals. Biology classes have long dealt with the fact that flowers are not random creations, but specifically designed to attract certain animals – insects, birds and bats. The fact that humans also find flowers attractive is a mere side-effect. Surprisingly, science did not begin to study the interrelationships between plants and animals systematically until the 1960s. A new concept developed: *Coevolution*. The term defines interactions which have occurred over the course of evolution between different organisms within an ecosystem. Coevolution is a continuous process of change which leads to the adaptation of one or both organisms.

Paul Ehrlich and Peter Raven, who introduced the term coevolution, believe that new plant species evolved from plants which began to produce and store poisonous substances in their leaves as a means of defense against being eaten. Tannins and other phenolics are found in the leaves of many rainforest plants and protect them from being eaten. The high diversity of plant species in tropical rainforests could thus be explained by the numerous different insect species present in

a warm climate. One might also assume that the insects adapted by becoming resistent to the plants' defenses and so themselves evolved into new species [89]. Whatever the case may be, the immense diversity of species in a rainforest environment allows for an equally immense number of interrelationships between plants and animals. Today, still little is known about these relationships and interactions. Our knowledge of the rainforest ecosystem is accordingly elementary.

The 1200 beetle species found on a single species of tree in Panama only hint at the complex web of relationships between insects and plants [53]. Even if not all the beetle species collected have a specific relationship with that particular tree species, many of them surely are associated with it in some manner. Research has shown that of the hundreds of fig species *(Ficus spp.)* throughout the world's rainforests, each is dependent upon one particular species of fig wasp for pollination. Of special interest are the relationships of plants with social insects, ants for example, since they occur in large numbers and may therefore have a stronger influence on plant evolution.

Plant – Ant Relationships

In comparison to beetle species, the diversity of ants appears to be ridiculously low. There are only 13 000 known ant species [90]. All true ants are social and live in colonies. Some of them have developed amazing relationships with plants over the course of coevolution. Researchers have long been interested in ants. Following an expedition by the American Museum of Natural History to the then Belgian colony of the Congo, a comprehensive publication on ants in African rainforests appeared in 1922. The work of over 1100 pages describes 90 genera of ants including over 300 species. The chapter on plant-ant relationships alone contains 250 pages [91]. The volume of this detailed information, the compilation of which certainly required much time and effort, makes us wonder if there is not actually more known about coevolution between plants and animals in the rainforest than is commonly assumed. In 1922, of course, the term "coevolution" and even today's common word "ecology" were unknown. Nevertheless, present-day researchers in these areas of science may still learn from past studies. Due to their remarkably early date of publication, however, they risk being disregarded.

Among other aspects, the researchers sent by the American Museum of Natural History paid particular attention to the relationship between ants and fungus, which is found in many ant nests. The ants in African rainforests, however, did not seem to intentionally cultivate fungi for food as do leaf-cutting ants (*Attini*) in the tropical and subtropical parts of the American continent. There, ants build their nests with freshly cut leaves and flowers and then inoculate them with certain fungi. The developing fungus mycelium decomposes the organic material and forms lumps at the ends of the fungal hyphae. The lumps are harvested by the ants as food and also serve to nourish the larvae. Surprisingly,

each species of leaf-cutting ant apparently cultivates its own fungus species.

But there have been others kinds of coevolution between fungi and ants in Africa. A fungus species in West and Central African rainforests has turned the tables and forced an ant to become its slave. Actually, the fungus makes cruel use of the ant to spread its spores. Although ants clean themselves thoroughly and often, the spores of this specialized fungus are able to implant themselves in the body of a large ponerine ant (*Paltothyreus tarsatus*). The ponerines are primitive ant species occurring in many African rainforests. They live in simple underground hollows and spend their entire lives on or under the ground. Equipped with a strong stinger, they hunt termites individually. Ponerines are also known as "stink-ants" because they produce a strange, musty odor. Once a microscopic fungus spore of the *Cordyceps myrmecophila* has implanted itself in the ant, fungal hyphae penetrate its body and eventually reach the insect's brain. The fungus causes the ant to change its behavior by programming its brain in such a way that it crawls up a bush or tree sapling, something it would normally not be inclined to do. The ant then bites firmly into the end of a twig – and dies. The fungus, however, lives on and continues to grow inside the ant's body. After some weeks, a stem of often more than two centimeters in length sprouts from the dead ant and develops a fruit body at the end. When ripe, it releases new spores to fall upon the next unfortunate ponerine passing by below (Fig. p.137). There are a number of other fungi which exist as parasites on living ants. Many of them are not choosy. They grow on various species of ants and sometimes on other insects as well [91]. The unusual aspect of the large ponerine's fate is that this parasitic fungus has specialized so precisely. By altering the ant's behavior, the fungus first uses its victim as a means of transportation to a strategically more advantageous posi-

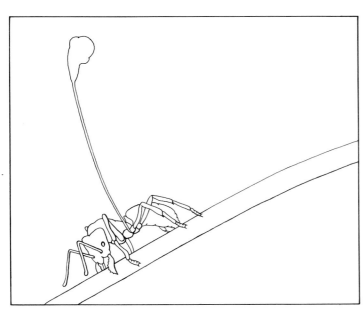

Forced into unnatural behavior by a fungus to finally be devoured by the same – the stiff body of a dead ponerine ant, *Paltothyreus tarsatus*.

tion for spore dispersal before killing the insect and feeding upon its body. It is not uncommon for a keen observer to find these poor little creatures with their fungus growths in the underbrush of the forest. In this case, coevolution has led to an interaction between two species in which one has become a parasite of the other. Phil Agland, who has probably made the best film ever on the ecology of African rainforests, succeeded in filming the strange fate of the large ponerine ant [92].

Ant trees

The older an umbrella tree *(Musanga cecropioides)* is, the more ants live on it. Besides a half dozen other ant species, ten different ant species belonging to the genus *Crematogaster* have been found to often inhabit this tree. One single tree can provide a home to some 200000 ants of this genus although it has no hollows suitable for hiding [93]. The *Crematogaster* genus is known for its ingenious card nests, which are hung in the tree branches. Although the umbrella tree is "friendly" to ants, no firm associations with specific species seem to have developed. Perhaps the umbrella tree is only an ecological niche of second choice where forest has been cleared. It is the most important pioneer plant in opened up areas of rainforest and corresponds in form and function to the South American *Cecropias*, trees which are closely associated with ants. But coevolution of plants and ants in West Africa also produced examples of close relationships between certain individual species. Symbiosis to the advantage of both participants is known as mutualism (as opposed to parasitism).

In Nigerian and Central African rainforests, local inhabitants avoid one small tree called "oko" in the language of the Yoruba. Scarcely reaching 15 meters in closed rainforest, the oko tree grows in the shadowy lower levels of the forest. It is the one tree spared when all others are axed down to clear land or free paths through the forest. The tree takes advantage of its special status and stretches its nearly horizontal branches wide. The magical powers of the oko's leaves and bark supposedly work even at a distance. They fend off black magic and turn the charm back upon the evildoer [94]. The power of the oko tree – scientifically known as *Barteria fistulosa* – has to do with its special means of defense: black ants *(Pachysima aethiops)*. The ants' sting is extremely painful and they readily attack anyone or anything threatening to harm the tree. The American ecologist, Daniel H. Janzen, studied the defense of the barteria tree in Nigeria and learned to respect *Pachysima* ants [95]: "When working under occupied *Barteria*, I became very preoccupied about being stung by *Pachysima*, something that has not happened with neotropical myrmecophyte ants... Within a few minutes, one or two *Pachy-*

sima workers found me either by falling from the tree or walking up from the ground; there is literally a slow rain of ants from a heavily occupied *Barteria.* A few seconds pass before a *Pachysima* sting begins to hurt, but when it does the pain is deep and throbbing and does not stop for one to two days. Janzen continues: "One to five *Pachysima* stings were enough to drive me away from a tree, leaving me very reluctant to return."

Once a young oko tree has reached a good meter in height, it begins to stretch out its first horizontal branches, which are hollow and

slightly swollen. Even trees of such small size may be colonized by a *Pachysima* queen. Young queen ants are winged and once a queen has alighted upon her chosen tree, she first checks to be sure it is an oko tree. She then gnaws her way into a hollow branch, lays her eggs and so founds a new ant colony. After about a year, the colony has grown large enough that the workers can begin to patrol the little tree. They gnaw openings all over the hollow branches and take up active defense of the tree. But they do not attack only humans and leaf-eating mammals. Caterpillars, beetles and other insects are also kept at a distance. A small colony of *Pachysima* ants may still have difficulty in fending off the larger weaver ants. But as time passes, the ant colony grows with the tree and slowly grows upward as the lower branches are shed. Janzen discovered that young oko trees occupied by colonies of *Pachysima* ants suffered noticeably less from herbivorous insects. The valiant ants also assure that no other plants get in the way of the scanty sunlight available. All other vegetation surrounding their host is gnawed to the ground, especially climbers since they could be of particular danger to the young oko tree. The leaves and twigs of other plants growing beside the crown of the oko tree are also pruned back. Its own leaves, however, are painstakingly cleaned of all organic particles which may fall from the upper levels of the forest.

The *Pachysima* colony continues to grow and eventually numbers 1000–4000 workers. They compensate the oko tree for the privilege of habitation by assuring it the best conditions for growth. They protect the tree and themselves at the same time. In order to defend the tree as best possible, the ants must be present at all times. They cannot afford to leave their posts in search of food, but neither can they eat from their own tree. Instead, an additional symbiosis has developed on the oko tree: A relationship between *Pachysima* ants and a species of clear-winged bug (*Homoptera*). The indolent bugs are kept and tended to by the ants inside the hollow branches of the oko tree. The *Pachysima* "milk" them of honeydew, a substance secreted by their anal glands. Since the bugs extract sap from the vessels of the tree, the *Pachysima* actually live off their host indirectly but without causing it harm.

The mutual relationship between *Barteria* and *Pachysima* is a rare example of a close plant-animal relationship which has been well studied. There are certainly many other cases of mutualism waiting to be discovered in Africa. Mutual

relationships between plants and animals, however, are seldom exclusive. Janzen found one in every hundred colonized oko trees to host a smaller related ant species (*Pachysima latifrons*) instead of *Pachysima aethiops*. It is dangerous to adapt so narrowly and thus become dependent upon a single species.

Seed Dispersal Systems

In 1977, the American scientist Stanley A. Temple published a much read article on the consequences resulting from the extinction of the dodo (*Raphus cucullatus*) on the island of Mauritius over 300 years ago [96].

The dodo probably evolved from a pigeon-like bird which had reached the island from the African continent millions of years ago. A very different habitat from that on the continent awaited the dodo's early ancestor: Strange plants, less competition for food and most importantly – no predators. Aside from a fruit bat, there are no mammals native to Mauritius. Neither are there fresh-water fish, nor amphibians. Isolated from the rest of the world, the modest pigeon evolved over thousands of generations to become quite an unseemly monster. Flight was no longer necessary since there were no enemies to flee from: The wings regressed to small stumps. The number of wing primaries decreased and the tail withered to a limp bush of feathers. The bird's weight and size, however, increased immensely. When Portuguese sailors landed on Mauritius in 1507, they were confronted with a pigeon the size of a turkey, its only means of defense being a powerful, hooked beak. The sailors chased the clumsy, flightless birds with clubs and caught them with their bare hands. In 1681, less than 200 years after its discovery, the fate of the dodo was sealed. Only a few engravings and one painted Indian miniature remain to bear witness to the existence of the legendary bird. The one museum specimen, which had been kept in Oxford, was mistakenly incinerated. Only a few parts could be salvaged.

Calvaria major, a majestic tree native to Mauritius, is represented by only 13 overaged specimens. Stanley Temple found that its seeds were unable to germinate. Although the surviving Calvaria trees continue to produce fruit, the seeds – embedded in a hard, woody shell – must have last germinated when there were still dodos. Temple concluded that the Calvaria tree must have coevolved with the dodo such that the seeds developed a 15-mm thick endocarp to prevent the nut from cracking in the bird's stomach. The thick shell, on the other hand, prevented germination if it had not been chafed in its intestine. Temple sought to prove his hypothesis by feeding Calvaria fruit to turkeys. After a period of up to six days of friction in the turkeys' stomachs, some of the nuts broke open but others passed through the intestine with chafed shells. They were deposited with the birds' dung – and germinated! The example of Calvaria major supports the theory that certain plant species have coevolved closely associated with a specific animal species and are thus irrevocably dependent upon it for good or for worse. Perhaps it was the enormous diversity of life forms within the rainforest, together with the incredible number of interrelationships between plants and animals, which led to the assumption that the fates of individual plant and animal species must also be closely related. In the short time since ecologists have begun to study the question of coevolution, they have at least often considered the Calvaria-dodo concept in their investigations.

Seed Dispersers in the Canopy

There never seems to be a lack of fruit in the rainforest. The majestic tree tops of the upper canopy bear fruit at almost every season of the year, if only on some branches. And when a gigantic emergent tree bears all at once, there appears to be an over-abundance and one wonders at the reasons for the plants' wasteful production. At least 50%, and often more than 75%, of the trees in a rainforest produce pulpy fruit – the moister the climate, the higher the percentage [97]. In West African rainforests, 70% of the taller trees and climbers bear pulpy fruit [28]. Plants which depend on wind for seed dispersal, however, only have a chance if they reach the upper canopy. These plants often need to time seed production with the dry season. Many seeds of this type are winged similar to those of the maple tree. In the lower levels of the forest, there are practically no wind-dispersed plants. There is not enough air current and the humidity is too high to allow light seeds suspended on a floss of hairs to be effectively dispersed. A similarly small proportion of wind-dispersed plants is found on isolated atolls, where conditions are likewise unfavorable for wind dispersal.

Fruit is meant to be eaten. Many ecologists have wondered if the high diversity of bird species in tropical forests does not mirror the abundance of fruit, an obvious connection when one considers the many fruit-eating bird species in West African forests. Hornbills are especially conspicuous birds, whistling through the tree tops and flapping their wings loudly. They live either solitary, in couples or in small flocks according to species and are among the most important consumers of fruit in the forest. Of course, they do not eat just any fruit but mostly pick out the small red, purple and black fruits with juicy pulp (Arillus) surrounding the seeds. They avoid the colors yellow, orange and green, leaving them instead to fruiting-eating bats, which also play a role in seed dispersal. With their over-sized beak, hornbills hurriedly gather fruit in the branches of the cardboard tree (Pycnanthus angolensis) and then fly to a thicket to devour their meal undisturbed. The eight or nine hornbill species which occur in West African rainforests often live sympatrically – they share the same patch of forest. Many tree species largely depend on hornbills to disperse their seeds which, after passing through the birds' intestine unharmed, are deposited in the soil with their droppings.

It would seem logical to categorize the various sizes of hornbills (Tab. 9) according to the size or species of the fruit they eat in order to prove a coevolution between certain species of tree and bird. But reality does not always conform to theory. Hornbills are opportunistic, they make use of what is available and feed on many different kinds of fruit; smaller hornbill species add insects to their diet. Often, several hornbill species will noisily dine together in the crown of a heavily fruit-laden tree. And they are not the only ones. Monkeys, too, will eat from the same tree. They stuff their cheeks with fruit and then withdraw to a secluded tangle of lianas to eat in peace. Like hornbills, some monkey species – cercopithecid monkeys in particular – are also important seed dispersers and compete directly with fruit-eating birds.

With so many creatures striving and scrambling for fruit in the upper canopy, it is hard to imagine how one plant could depend on a single animal species for seed dispersal as was apparently the case with the *Calvaria* tree in Mauritius. This was confirmed by a recent field study conducted in a lowland rainforest area of Gabon [98]. A team of scientists investigated the interrelationships between fruit-eating vertebrates and fruit-bearing plants over the course of one year in the M'passa Reserve near Makokou. For example, 80 kinds of fruit were identified as being eaten by either birds or monkeys. The animals often also dispersed the seeds. 42% of the fruit species were eaten by both groups, which means their diets overlapped to a great extent and brings up the question of competition. The scientists described the typical appearance of fruit consumed by tree-dwelling vertebrates as the "bird-monkey syndrome": A colorful, sweet, juicy berry containing one or more seeds with thin walls. In this case, the individual plant must adapt its seed dispersal system to satisfy a hord of greedy and opportunistic takers rather than a single specialized species.

Seed Dispersal by Ungulates and Rodents

A hiker through the rainforest will soon notice quite a different abundance of fruit once he finds himself suddenly standing upon a soft, bitter-sweet smelling carpet 10, 20 or more meters wide. But beneath his feet are not the colorful berries made to attract birds and monkeys; despite their good size, these fruits are difficult to discern on the shadowy floor of the forest due to their modest green or brown color. Embedded in the sticky, fibrous pulp is nearly always a hard, woody pit in which the seeds are well protected. Slightly unappetizing in appearance, this fruit is apparently designed to attract very different customers. Rodents and ungulates serve to disperse the seeds. Since hardly any

Table 9
Species diversity and maximum wing length of male hornbills in the Bia area of Ghana [63,66].

		Wing Length (mm)
Red-billed dwarf hornbill	(Tockus camurus)	157
Black dwarf hornbill	(Tockus hartlaubi)	166
White-crested hornbill	(Tropicranus albocristatus)	243
Piping hornbill	(Bycanistes fistulator)	257
Allied hornbill	(Tockus semifasciatus)	265
Brown-cheeked hornbill	(Bycanistes cylindricus)	323
Black- and white-casqued h. bill	(Byanistes subcylindricus)	337
Black-casqued hornbill	(Ceratogymna atrata)	408
Yellow-casqued hornbill	(Ceratogymna elata)	410

grass and few edible herbs grow on the forest floor, ground-dwelling mammals take advantage of the fallen fruit. Duikers, the bushbuck and the bongo consume a great deal of fruit but usually destroy the pits and their seeds. Some of the pits, however, are spit out during cud-chewing and others pass through the intestines of the larger ruminants unharmed. Seed dispersal by ungulates is not very efficient, however. The bush-pig and the giant forest hog can crack even hard nuts with their powerful jaws, thus eliminating the plant's chances of reproduction. Only the elephant is known to disperse large numbers of seeds with intact shells and may transport them over long distances before depositing them with its dung.

There is no clear difference between seed dispersers and seed predators. Plants run the risk of losing their seeds by tempting fruit-eaters with their nourishing fruit pulp. Often enough, an animal will eat a fruit and the seed to its own advantage only and the mutualism of the relationship is forfeited. But even after a seed has passed unharmed through the jaws and intestines of an ungulate, it is by no means guaranteed to germinate. And should it find its way into fertile soil and receive sufficient sunlight, the risky business is still not finished, for seed thieves lurk in the shadows: mice, rats, squirrels and other rodents. They are less interested in the fruit pulp and gladly allow hoofed animals to take care of the hindrance. If life were a bowl of cherries, rodents would be happy with the pits. They immediately begin to gnaw their way to the tasty seed awaiting them inside. Old elephant dung is often full of nuts, or pits, which have been opened by rodents. The seeds are gone. But certain rodents may also act as seed dispersers: Giant rats hoard large quantities of fruit and nuts in their underground dens. Occasionally, some of them germinate before they can be eaten. The brush-tailed porcupine hides away large quantities of fruit and nuts beneath fallen wood and clearly contributes to the dispersal of certain seeds [98].

The best possible way for a plant to prevent all too much damage to seeds dispersed by ungulates and rodents is by investing as much energy as possible in a hard, protective seed covering. But a thick endocarp – the botanical name for the hard shell – can be dangerous, as seen in the case of the *Calvaria* tree. If the pit becomes too resistant, the germ inside may never see sunlight.

Coevolution between Plants and Seed Dispersers?

There is no doubt that pulpy fruits are an adaptation to seed dispersal by animals. But what kind of fruit best guarantees success? What kind of fruit will attract animals likely to disperse the seed without harm? Doyle McKey, who studied seed dispersal systems in the rainforests of Cameroon, assumes that most fruit-eating animals do not disperse seeds very efficiently. The better a plant adapts its fruit to the most efficient seed dispersers, the better its chances of survival. McKey postulates that selection on the part of seed dispersing animals influences the evolution of fruit in various ways [99]:

1. Adaptation to as many seed dispersers as possible. The seeds must be small enough to be transported by many animal species. Large quantities of seeds are necessary, however, since losses are high.

2. Adaptation to a specific group of animals, which often feed on the particular fruit and disperse fewer seeds more efficiently. The fruit must be nourishing to be attractive enough not to be passed over by the relatively small number of seed dispersers.

3. No seed dispersal by animals. Instead, seeds are adapted for dispersal by wind, water or even by the plant itself (seed ejection). Fruit

The fruit of the liana *Strychnos aculeata* is as hard as a rock. Elephants alone are capable of cracking its shell with their strong jaws. *Strychnos* seeds can be found germinating in elephant dung all year round.

pulp is not required but dispersal in the rainforest is limited to a relatively short distance.

Plant species can use still other methods to improve their chances of seed dispersal: Individual plants of one species may bear fruit at different times of the year in order to reduce competition among themselves. This is the case with many tree species in the rainforest. Individual trees bear fruit at different seasons and fruit-eating animals are attracted from tree to tree across the forest and throughout the year. Improved seed dispersal has been said to be the reason for the varied fruiting seasons of rainforest trees [100].

It would seem that coevolution does in fact occur between plants and animals with respect to seed dispersal. But it is incorrect to assume that a plant in an environment as rich as the rainforest adapts to one specific seed disperser. Or in other words: A higher animal species is not likely to depend on a single plant species for its food supply. In Gabon's lowland rainforest, 122 plant species with pulpy fruits were checked on their seed dispersal agents. Approximately half of them were found to be consumed and partially dispersed by at least three different groups of fruit-eating animals [98]. Coevolution of seed dispersal systems thus means interrelationships between groups of animal species and the plants they feed upon – interrelationships which vary from place to place and from time to time. Or is there another exception to the rule?

Forest Elephants and Seed Dispersal

What appear to be yellow cannonballs lying in the middle of the rainforest are actually a kind of fruit – a seemingly strange fruit. Although similar in appearance to round pumpkins, this fruit is as hard as a rock and just about impossible to crack open. Someone unfamiliar with African rainforests may wonder where these things come from. A glance above affords no answer. The next surprise awaits and the newcomer tends to place the fruit in the category of puzzling phenomena, as is the case with so many aspects of the rainforest. But after one of those cannonballs has come crashing down from above, ripping its way through the leaves to land at the newcomer's feet, second thoughts are likely to occur. The dangerous object originates from above after all, although no tree with yellow cannonball fruit can be sighted. They are not to be found in large quantities on the ground – a dozen here, a few there. And surprisingly enough, they seem to occur all year round. Whatever plant produces this strange fruit, it has a special system of seed dispersal. It encloses its seeds in a large shell as hard as a rock and distributes it sparingly throughout the year. The plant in question is a large, spiny liana (*Strychnos aculeata*) which climbs its way through the trees to the upper canopy where it bears at first green-colored fruit almost impossible to discern from the ground below. The fruit shell, which is approximately seven millimeters thick, contains about 20 flat, round seeds arranged like the spokes of a wheel and embedded in soapy slime. In the Côte d'Ivoire, the slime is used to stun fish. Elsewhere, the crushed seeds are used to induce vomiting. As suggested by its scientific name, the liana contains poisonous substances in its stem and bark. In small quantities, they are used for medicinal purposes.

It is an open question whether or not the strychnine fruit also has a doping or intoxicating effect on elephants. Certainly, elephants throughout West African rainforests eat the fruit and do a good job of dispersing the seeds: *Strychnos* seeds are often found in elephant dung, where they soon germinate. In spite of what ecologists have assumed, has this liana coevolved its seed dispersal system to depend on the elephant alone? It would actually be difficult to imagine another animal capable of eating or merely breaking open this difficult fruit. Only the jaw of an elephant is capable of such a task. Even those scientists whose studies showed many groups of animals responsible for seed dispersal in Gabon's rainforests admit that in some cases the elephant seems to be the only answer. Perhaps the plant kingdom has developed special adaptations to accommodate the largest animal on the continent. Whatever the case may be, it is worth taking a closer look at the role of the elephant.

Forest Elephants or Pygmy Elephants?

There is only one species of elephant in Africa: *Loxodonta africana*. Today, scientists more or less agree that there are two subspecies, the savannah or bush elephant (*L. a. africana*) and the forest elephant (*L. a. cyclotis*). It was 70 years, however, before this consensus could be reached. In 1975, one German scientist still insisted on having been on the track of a separate species of pygmy elephant in southwestern Cameroon. His reports led the authors of an otherwise very useful identification manual on African mammals to include a pygmy elephant (*Loxodonta pumilio*) and declare it the least known large mammal on the continent [101]. That was probably neither the last, nor the only confusion the forest elephant has caused among taxonomists. Günter Merz, author of a dissertation on

That scientific knowledge depends on the possibility for observation is clearly shown by the example of forest elephants. Very few studies have been made on the habits of this forest inhabitant often hidden deep within the forest although nearly half of the African elephant population today consists of forest elephants. The interior of Central Africa's rainforests provide them with a haven still largely safe from poachers, a problem which has become very severe for the savannah elephant.

Traditional elephant trap along an elephant trail in the rainforests of Central Africa. Such traps are hardly used in West Africa today.

forest elephants, however, leaves little doubt as to classifications based on hearsay and assumptive single records. During 1978–80 he collected large amounts of field data on forest elephants in Taï National Park in the western Côte d'Ivoire. His work cleared up much of the confusion caused by scientists who had never seen a living specimen of forest elephant [102].

The question of the difference between elephants in the rainforest and those in the savannah was not about the fact that the forest form is of lesser size, has somewhat smaller, more rounded ears; that its forehead is flat and its tusks longer and thinner. Scientists had long agreed on this. The difficult question was just how small were these mysterious creatures of the forest? Indeed, it is a difficult question to answer. Forest elephants have a body structure similar to that of young savannah elephants. After glimpsing an elephant in the undergrowth, one can hardly be certain if it was actually a forest elephant or perhaps only a young savannah elephant. The age of an animal is another crucial question: A bull, at least, continues to grow practically all its life and there is no definitive shoulder height for adult animals. This makes a size estimate of an individual animal useful only if its age can also be estimated. And finally, there is the difficulty of observing forest elephants. Whoever is lucky enough to spy a forest elephant with his own eyes, must satisfy himself with a few seconds of gray skin or a bit of white tusk shimmering behind the leaves. Unless, of course, the observation is sudden and at an unpleasantly short distance, in which case the shocked observer will easily lose his sense of proportion. Not surprisingly, the animal has seldom been photographed. The only reason for the long uncertainty about the forest elephant is the fact that it is such a difficult subject to observe and the rainforest is such a difficult environment for field work. Meanwhile, wildlife biologists had long collected mounds of data on the savannah elephant in eastern and southern Africa. There is no lack of publications on the larger subspecies.

Varying Populations of Forest Elephants

Just how large are forest elephants really? A generally accepted rule defines an elephant herd as being forest elephants when the average shoulder height among all the adult animals measures less than 240 centimeters [103]. But how are such measurements to be taken? Various researchers have made use of a simple rule concerning the body proportions of elephants. Twice the circumference of a standing elephant's forefoot is approximately equal to its shoulder height [104, 105]. Forest elephants move through the forest alone or in small groups of up to four animals. Larger groups of up to about ten individuals are relatively rare. The animals often leave clear tracks on soft ground and individual footprints are usually easy to measure.

In 1977/78, two different West African forest elephant populations were studied [66, 106, 107]. The footprints of both populations were measured to attain a size distribution of each group (Fig. p.147). The largest forefoot circumference measured among the forest elephant population of Taï National Park was 110 centimeters, which leads to a calculated shoulder height of only 220 centimeters. The shoulder height of 97% of the animals measured less than two meters [102]. On the other hand, about half the forest elephant population of Bia National Park and the surrounding areas was calculated to have a shoulder height of above two meters. The largest forefoot circumference measured 142 centimeters, which indicates a bull of over 2.8 meters at the shoulder. That is still approximately one meter less than the height a savannah elephant bull may reach. Bia National Park lies only about 100 kilometers

south of savannah territory and the elephants there have to be considered an intermediate form of the two subspecies. Thus, there is no standard size for forest elephants and the average shoulder height may vary from herd to herd. Neither can a clear line be drawn between the distribution ranges of forest and savannah elephants. In the transitional zone between rainforest and savannah, a variety of intermediate forms of elephants can therefore be recorded. And sometimes it is simply impossible to clearly classify a population or herd of elephants under one of the two subspecies. This is, however, no reason not to distinguish two different subspecies of African elephants. Nevertheless, neither

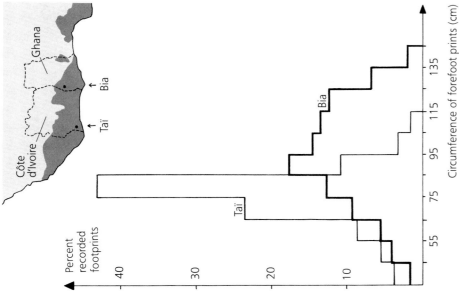

Forefoot circumferences of forest elephant footprints measured in Taï National Park [102] and Bia National Park [66].

does a too rigid classification provide an adequate solution to the mystery of Africa's forest elephants.

The smallest forest elephant populations occur in the wettest areas far from savannah regions. They are found on the west coast of Africa and in the interior of the Congo Basin. It is here that the largest continuous distribution range of forest elephants remains. Population estimates for elephants made in 1987 list 332000 individuals in Zaïre, Gabon and the Congo (Brazzaville) alone [108]. By far, the majority of these elephants may be classified under the forest subspecies. In West African countries, however, elephants have suffered severe losses and only small, isolated populations remain (Tab. 10). In the early 1980s, 35 to 40 more or less isolated populations survived in the Côte d'Ivoire. The smallest consisted of only a few individuals [109]. Forest elephants no longer inhabit large areas of the West African rainforest zone, a fact which may affect not only the elephant but other species as well.

Table 10
Country estimates of elephant populations along the Gulf of Guinea in 1987 [108].

Benin	2 100
Ghana	1 100
Guinea	300
Guinea Bissau	20
Côte d'Ivoire	3 300
Liberia	650
Nigeria	3 100
Sierra Leone	250
Togo	100

Note.: The above estimates include both subspecies of the African elephant. The figures for Benin and Nigeria most likely include savannah elephants only. In Ghana and the Côte d'Ivoire, at least half of the numbers listed are likely to consist of savannah elephants (L.a.africana). The population of forest elephants in West Africa is thereby reduced to no more than 3000 animals.

Opportunistic Feeding Patterns

Water holes are not the focal point of elephant activity in the rainforest as is the case in the drier regions of the continent. Water is readily available and water loss through transpiration is counteracted by a relative humidity of about 90%. Even so, forest elephant groups do not wander aimlessly through the forest. Günter Merz discovered that elephants in Taï National Park have a special preference for gaps in the forest thickly overgrown with secondary vegetation. Many of the fast growing pioneer plants common to such habitat have a high protein content and Merz assumed this to attract elephants. Elephant dung counts in sample areas, as they were also conducted in Bia National Park, led Merz to calculate surprisingly high population densities of 1.08 to 2.6 individuals per square kilometer in some areas outside the primary forests of Taï National Park. Within the park's area of 3300 square kilometers, however, the average elephant population density in 1980 was found to be only 0.23 animals per square kilometer [102, 110]. Elephants also congregated in areas of secondary vegetation inside Bia National Park [106]: The leaves and twigs of 138 plant species could clearly be identified as "elephant food" and that is not even a complete list of what elephants rip down and break off on their travels, especially through secondary forest (Tab. 11). The majority of those

Table 11
Number of plant species of which forest elephants in Bia National Park consume leaves and twigs [106].

Type of Plant	Number of Species
Trees taller than 8 m	48
Trees/shrubs smaller than 8 m	30
Large, woody lianas	47
Smaller climbers	12
Ground herbs	1

Elephants traditionally follow the same trails and do not meander aimlessly through closed rainforest (photo right). Although elephants repeatedly strip pieces of bark from certain trees, they seldom do fatal damage. A healed scar is visible on the trunk of this tree just below the guard's hand.

plant species are climbers often found at the edge of small gaps and clearings. Elephants feed on 57% of the larger liana species found in Bia National Park. And as if wishing to round off their leafy diet, they also consume bits of bark from some 20 different species of trees, which they loosen with their tusks and pull off with their trunks. The bark is usually spit out after being chewed down to a bundle of fibers. But tree trunks in the rainforest are much too large that the elephants could do lasting damage to them, as is often the case in savannah regions.

Thus, the forest elephant is anything but choosy in its diet. It takes advantage of a great variety of rainforest vegetation in one way or the other and interestingly enough, different populations seem to have different tastes. In Taï forest, elephants commonly feed on the leaves of *Funtumia elastica.* But in Bia National Park 400 kilometers to the east, elephants pass up the tree altogether. And the Bia elephants' favorite climber, *Microdesmis puberula,* does not appeal to elephants in Taï. Such food preferences are surely also related to the local abundance of certain plants, but they indicate that the elephant does

Minutes before this photo was taken, a group of forest elephants crossed the narrow Hana River inside Taï National Park. In the rainforest, one learns to make do with tracks.

not necessarily follow a strict diet. Its use of the forest vegetation is opportunistic rather than specialized and it is often only by chance that smaller fruits and seeds find their way into its stomach accompanying large masses of leaves and twigs. Seeds which pass through the animal's intestine unharmed are deposited in the soil with its dung for germination. 93% of the elephant droppings examined in Bia National Park contained seeds or at least bits of fruit. But the majority of those seeds are the large, hard pits of pulpy fruits which were not eaten by chance. Elephants will journey long distances, alone or in small groups, in search of very specific fruits. Such journeys lead them far from the sunny clearings deep into closed forest.

Elephant Paths Lead to Fruit Trees

One of the greatest known forest botanists, A. Aubréville, reported in 1958 that seedlings of the makore tree *(Tieghemella heckelii)* were often strewn along the paths followed by elephants [111]. Botanists apparently also made occasional use of elephant trails long known to local hunters and gatherers crisscrossing the forest undergrowth. Although the trails do not follow straight lines, neither do they meander aimlessly. Besides, it is easier to move along elephant trails than along paths cleared with bush knives. The sharp ends of cut branches and small trunks often hinder progress along man-made paths. Elephants use the same routes for years. A star of trails will converge at specific places where the forest undergrowth has become thinned out in catacomb fashion. At these sites, tree seedlings are repeated trampled and vines are continuously ripped down, for the elephants' true interest is held by the tree in the middle of these mysterious meeting places.
In southwestern Ghana, elephant trails through the forest often converge around a majestic makore tree which may reach 60 meters in

height and have a diameter of up to three meters. Elephants favor the tree for its greenish-yellow fruit, timber companies are interested in its red hardwood, a commodity in high demand. Another fruit tree preferred by elephants, *Parinari excelsa,* is of slightly smaller size. When there is little else to eat, local human inhabitants apparently also harvest the fruit of *Parinari,* otherwise known as the Guinea plum tree. Middle-sized trees can also be favorites among elephants: *Balanites wilsoniana,* with its fluted trunk and *Panda oleosa,* which reaches a height of only 20 meters, are other attractions. All four tree species bear somewhat dull-colored, pitted fruits five or more centimeters in diameter. The fruits, which can be very pungent or even unpleasant in smell, all enclose a large, nutlike pit (Fig. p.152). They are capable of passing through an elephant intestine unharmed and are deposited by the dozens with the animals' dung throughout the forest where they soon germinate.
Another French botanist, D. Y. Alexandre, studied the early germination of forest tree seeds in elephant droppings. He discovered and identified 48 tree species in Taï National Park which germinated in elephant dung. Many of them were among the tallest trees of the forest but they were not the only plants to sprout from elephant droppings: Seedlings of the *Strychnos* liana were also common [112]. A small, very rare species of tree *(Pancovia turbinata),* which had never been observed to bear fruit locally, was also discovered to germinate in the droppings. Not only can large nuts survive the journey through an elephant's digestive system, small seeds are also deposited for germination. Alexandre gathered a variety of seeds from fresh elephant dung and compared their germination with that of seeds he extracted from the fruit pulp himself. In the tree nursery, however, only few species showed differences. Successful germination was rather modest both for the seeds

The thick, woody shells of these pits protect the seeds on their journey through the elephant's intestine so that they can later germinate when deposited with dung. These pits ("nuts") are depicted in original size and are quite often distributed by elephants.

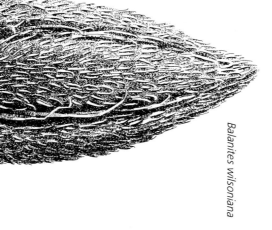

Balanites wilsoniana

Parinari excelsa

Tieghemella heckelii

Panda oleosa

Trampled paths from all directions converge at this opening in the undergrowth of the forest. Forest elephants congregate here when the fruit of the tree (*Panda oleosa*) in the middle ripens and falls.

Seedlings in decaying elephant dung.

from elephant dung as well as for those extracted from the fruit. In contrast, seeds which remained embedded in dung grew within three months to healthy seedlings with well-developed root systems. Experiments conducted in Ghana showed similar results: Very few of the seeds with thick coverings collected from elephant dung germinated more successfully in pots [113]. It seemed that passage through an elephant intestine was less important for germination than the fact of a seed's being planted in dung.

Fruit Attracts Elephants

From December to March during the dry season, the Harmattan winds blow across the rainforest zone bringing dust from the desert. Some of the tallest trees in the moist semi-deciduous forest zone shed their leaves and in Bia National Park many of the *Panda* and *Parinari* trees bear fruit. There is the sharply pungent fruit of the *Balanites* tree as well as the "wild mango" (*Irvingia gabonensis*). Increasing numbers of elephants are attracted to the park from the south and the population density rises to an average of 0.44 animals per square kilometer [107]. The Guinea plum tree (*Parinari excelsa*) seems to be especially attractive. During the dry season of 1977 in Bia National Park, few people dared to leave the research camp located in the park's center. Elephants congregated for weeks around the research station to feed on the fruit of the *Parinari* tree common there.

In April and May, when more rain begins to fall and in the major rainy season following, few fruits are found in the forest. In the park, the population density falls to an average of 0.13 elephants per square kilometer. The animals are distributed over a wide region; they frequent secondary forests and plantations in the park's surrounding areas. The wettest months of the year, from April to July, are when elephants do the most damage to crops. During the shorter minor rainy season from September to November, elephant presence increases south of the park in the neighboring Bia Wildlife Reserve and in the surrounding forests. This is the season when the numerous makore trees let their jumbo-sized fruit fall to the ground. Merz found similar concentrations of elephants around two further species of trees in Taï National Park [102]: *Duboscia viridiflora* and *Sacoglottis gabonensis*. The latter often grows at water's edge but although it is a timber species, it is scarcely used commercially. The fruit of the *Sacoglottis* is favored by elephants. It has a high resin content and can stay afloat for a long time.

The elephants in Bia and Taï prove that certain tree species influence elephant activity. When large quantities of pitted fruit fall from these trees, they attract forest elephants and thereby lead to concentrations and temporarily increased population densities. Other fruits are eaten more by chance because they are either less attractive or not abundant enough to induce the animals to migrate. A third category consists of fruits like those of the *Strychnos* liana which are available throughout the year, but never in large quantities (Tab. 12). In spite of the fact that a forest elephant consumes a wide spectrum of plants and is not confined to a narrow diet, it does have its preferences and plays an important role in the seed dispersal of certain tree species.

Incidents on Plantations

Torn by their preferences for secondary vegetation and the fruits growing in primary forest, forest elephants often commute between the two. In the Bia region, elephants often cross the park border into the surrounding opened forest areas where they deposit their dung filled with seeds and so contribute to the regeneration of

the vegetation. They contribute to the dispersal of closed forest tree species into forest fallow areas. In excursions of one to three days, elephant groups will wander through secondary vegetation and plantations before returning along their traditional trails to the fruit trees in the park (Fig. p.155). But their excursions into plantations do not go unnoticed by the farmers.

Table 12
Fruits of which seeds were found in Bia-elephant dung.
The figures refer to the proportion (in %) of dung piles in which seeds of the various species were found at different times of the year. Elephants are especially attracted by the underlined species. a: trees taller than 8 meters, b: trees and shrubs smaller than 8 meters, c: large lianas, d: smaller climbers.

Species	Dry Season	Main Rainy Season	Small Rainy Season
Strychnos aculeata (c)	41	66	35
Myrianthus arboreus (a)	2	54	32
Ricinodendron heudeloti (a)	1	2	15
Antrocaryon micraster (a)	15	2	37
Duboscia viridiflora (a)		2	10
Balanites wilsoniana (a)	11	10	
Lagenaria breviflora (d)	13	10	
Klainedoxa gabonensis (a)		32	
Vitex ferruginea (a)	13		32
Grewia malacocarpa (c)	2		15
Desplatsia sp. (b)	11		7
Tetrapleura tetraptera (a)	65		5
Tieghemella heckelii (a)			12
Parinari excelsa (a)	66		
Panda oleosa (a)	34		
Irvingia gabonensis (a)	26		
Chrysophyllum beguei (a)	17		
Craterispermum caudatum (b)	11		
Uapaca guineensis (a)	9		
Theobrama cacao*	4		
Strombosia glaucescens (a)	4		
Omphalocarpum sp. (a)	4		
Telfairia occidentalis (d)	3		
Treculia africana (a)	3		
Buchholzia coriacea (a)	2		
Chytranthus carneus (b)	2		
Swartzia fistuloides (a)	1		
Carica papaya*	1		
No seeds found	2	16	15

* Seeds of crops cultivated on nearby plantations.
Source: Short [106].

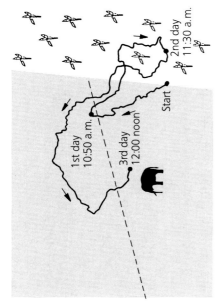

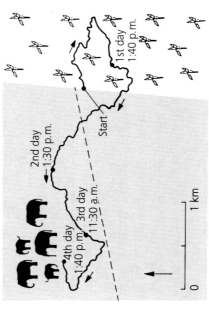

Forest elephants tend to "commute" between closed rainforest (left) and secondary vegetation or plantations (right) without covering very long distances. The diagrams show the respective routes taken by a solitary bull and by a group of females with offspring. The routes were recorded in September of 1978 on the border of Bia National Park.

The elephant's inclination to regularly feed on plantation crops in opened up forest areas often leads to conflicts which usually end in the animal's misfortune.

In the areas surrounding forest reserves in Ghana, game wardens are often obliged to come to the aid of farmers. The remaining elephant populations are increasingly crowded in by development outside the park and need to be "controlled". In other words, the wardens must shoot animals responsible for crop damage. But the land conflict is not always solved to the farmer's advantage. In June 1977, an armed game guard was called to a plantation at the edge of the Bonsan Bepo Forest Reserve. Among the plantains stood an aged elephant bull. Large, charred trunks scattered everywhere attested to the fact that fire had been recently used to clear the land. Cocoa tree seedlings had been planted between the bananas to establish a future cocoa plantation. The guard ordered the three farm workers who had called him to step back, he loaded his gun and fired two lateral shots at the bull's chest. The seriously wounded animal sank to its knees and – rose again. As it slowly began to approach the small group of men, the guard reloaded his gun and motioned the others to retreat. One of the farmhands lost his rubber sandal fabricated from an old tire and without thinking turned around to grab the shoe – it was to cost him his life. The fatally wounded elephant attacked, cut off the man's route of escape and crushed him against a nearby fallen tree trunk. One of its tusks penetrated the right side of the poor man's chest and he died with a gasp of despair, "Wa ku mi oh" (he has killed me – oh). In this case, forest, elephant and man were sacrificed in a land conflict. Similar incidents occur in other places where elephant populations have been isolated in too small patches of forest.

No Elephants – No Makore

Whereas Aubréville had found ample seedlings of makore (*Tieghemella heckelii*) along elephant trails in the 1950s, botanists Hall and Swaine failed to find young specimens of the tree in Ghana in the 1970s. They concluded that this finding was the result of a lack of elephants [28]. The makore tree occurs only in the rainforests of eastern Liberia, the Côte d'Ivoire and Ghana where, until a few decades ago, elephants were widely distributed. It seems probable that coevolution has led to the tree's dependency on the elephant for seed dispersal. Botanist Alexandre is less reserved in his conclusions. He, who was the first to systematically study seed dispersal by elephants in the Taï region, concluded that the animals' fruit-eating was not a "random" phenomenon, but an "essential" activity for the trees. He presumed the elephant to be by far the most important seed disperser for nearly 30% of the larger tree species and over 40% of those in the upper canopy. Long-term floristic diversity thus depends on the elephant. If one accepts Alexandre's findings, and there is little reason not to, it would follow that other species besides the makore tree must have also become rarer. Until today, however, forest inventories have not paid as much attention to other species as they have to the makore since they are not of commercial value.

Whereever larger populations of forest elephants exist in West Africa, there is still primary forest as well. In spite of its preference for secondary vegetation, the elephant has remained dependent on closed rainforest. Closed forest is vital to its survival, less because of the fruits found there and more due to the land conflicts with farmers who defend their land with arms. Illegal hunting is a general problem. But neither have elephants in protected areas escaped poaching. In the course of his study, Merz noted a sharp drop in the Taï elephant population from about 1800 to an estimated 800 individuals in 1980. At the start of 1988, there were perhaps only 100 forest elephants in Taï National Park [114], the largest protected area in West Africa! More recent estimates of the Bia elephant population in Ghana, which consisted of 200–300 animals in 1978, have not been made. Doubtlessly, this population, too, has suffered from hunting pressure.

The days of the forest elephant in West Africa are numbered if forest destruction and poaching continue over the next few years. The theoretical question of whether certain plant species have coevolved their seed dispersal systems to depend solely or to a great extent on the elephant will become rather incidental. Nevertheless, the question deserves an answer. What other animal, what other selective factor could have caused the evolution of fruits with pits of four to five centimeters in diameter and shells six millimeters thick, if not the largest land mammal? Is it mere coincidence that the elephant of all animals is crazily attracted by the smell of those same fruits and that it will journey long distances in search of them? Such behavior corresponds to a plant's adaptation to a specific seed disperser as postulated by McKey in his list of three strategies [99]. Clearly, certain African tree species bearing large pitted fruits have coevolved their seed dispersal systems to depend almost solely upon the elephant.

Forest elephants take revenge for slash-and-burn clearing. They feed on the stalks of banana plants – much to the displeasure of the farmer.

The Forest as Human Habitat

Although mankind's earliest origins may be traced back to African rainforests, our Stone Age ancestors had long ceased to favor the rainforest as a place to live [1]. The climate fluctuations from cool, dry periods to warm, wet spans and the resulting retreat and expansion of the rainforest zone during the Pleistocene also affected human settlement. Dry periods led to migration from the Sahara to the bend of the Niger River and from the savannah zone into the rainforest. Supported by archeological findings, such migrations lead us to assume that prehistoric man was better suited to life in the savannah than in the desert or in closed rainforest. During the Middle Stone Age there were a number of migrations from the Niger Valley, not far from today's Niamey, to the Volta region of Ghana up to the Accra plains and to the edge of the rainforest zone but no further, at least not at that time.

It was not until the Late Stone Age, after 900 B.C., that two groups of settlers pushed into the rainforest zone from the east and northeast. They reached the area of Conakry in the Republic of Guinea. Somewhat later, the Kintampo culture introduced the art of pottery from the Niger Valley; stone axes, picks, arrowheads and other stone tools found more and more use inside the rainforest. But the population density was low, the settlements were scattered along the rivers. Even then, these Neolithic settlers already cultivated the soil. Yams (*Dioscorea spp.*) of which there are many wild species in the savannah and in the rainforest, were planted and harvested in crude plots of land. Agriculture, however, played a minor role for the early inhabitants of the rainforest. They did not yet know of slash-and-burn farming methods and lived mostly from hunting and gathering in the forest as the pygmies in Central Africa still do today.

There were no signs of shifting agriculture in the rainforest zone until 500–1000 A.D. when tribes that had knowledge of the art of working iron intruded from the north. The art stemmed from the Nile Valley. Iron, however, did not completely replace stone in fabricating tools in the rainforest until just a few centuries ago. Settlements were not permanent but changed according to the principles of shifting agriculture and were usually located in the vicinity of rivers. It was then that basic tribal structures began to develop. Forest-dwelling people from the Côte d'Ivoire, through Ghana, Togo, Benin, up to western Nigeria are related as is evident by their languages – they all belong to the linguistic group Kwa.

Around the 16th century, Portuguese travellers introduced new crops from Latin America and Asia to the west coast of Africa: Sugar cane, pineapples and bananas, sweet potatoes, oranges, limes and red peppers. Just when corn was introduced is uncertain. Plantains were possibly introduced much earlier from Southeast Asia. There is proof, however, that the Portuguese brought doves, chickens, pigs and sheep with them. Increased agriculture led to higher populations in certain areas of forest [1]. While populations in southwestern Nigeria was relatively heavily populated by the Yorubas from the 14th to the 17th century, wide areas of forest in West Africa remained practically uninhabited by man until this century [115]. Hunters played an important

role in forest settlement. Everywhere in West Africa, one hears stories of how hunting camps 10–30 kilometers from the nearest settlement often developed into settlements of their own. The name "Sunyani", today a small city in the Brong-Ahafo Region of Ghana, indicates a place where elephants were skinned [1].

Thus, agriculture is a relatively new tradition in West African rainforests. And should some forest peoples have used fire to clear land in pre-Christian times, the damage to the forest could not have been lasting. There is certainly no reason to believe that a few pre-Christian settlers could have caused the limited plant diversity of Africa's rainforests as has been suggested [34]. In thinly populated areas, even land cleared by fire only a few hundred years ago has regenerated to such an extent that it is scarcely distinguishable from primary rainforest.

"Secondary" Forest Utilization

For centuries, forest-dwelling people have lived by hunting and gathering produce from the forest. Agriculture was a minor addition to their food supply. Today still, the culture and economy of forest-dwellers are a balance of hunting, gathering and agriculture. Knowledge of plants and animals, their uses for food, medicine or for other purposes was passed by word of mouth from one generation to the next. Man had lived long enough in the forests of West Africa to develop intricate utilization patterns. Local handicraft developed and raw material gathered from the forest was used in various ways. The Akan people, of which the Ashantis in Ghana are the most dominant, actually made textiles from rainforest plants. Long pieces of bark were ripped off Kyenkyen trees (*Antiaris toxicaria*), soaked in water and beaten with wooden clubs. The result was a soft piece of cloth much wider than the original piece of bark. The production of "bark cloth" was an important part of the local

Blue smoke seeps through the wet straw roofs of kitchens in Toko, a forest village in southwestern Cameroon; bushmeat is being smoked. As long as the people here can remember, they have always lived off the forest without destroying it.

economy in the rainforest zone. In the mean time, the Ashantis had introduced the art of Kente weaving. Kente cloth is sewn of narrow strips woven in traditional patterns. Kente cloth and European textiles eventually replaced "bark cloth". But until this century, hunters and poorer people in the Brong-Ahafo Region of Ghana still wore clothes made of Kyenkyen cloth. Today, the forest continues to be a source of fruits, vegetables, nuts, oils, spices, tannins, fibers, resins, rubber and medicinal substances as well as bush meat, skins, honey, firewood and building materials. The official term used by for-

esters for all of the above is "secondary forest products", their use is categorized as secondary. The term also reveals their status. For some decades now, the primary forest product – commercial timber – has been considered more important, usually at the cost of all other less damaging kinds of forest utilization. The socio-economic value of hunting and gathering in intact rainforests has been grossly ignored in West Africa and elsewhere. Timber exploitation now opens up the forest to an invasion of farmers from the north. Age-old knowledge and traditional uses of the forest are likely to be lost.

Even better than bags made of recycled paper, *Marantaceae* leaves from the rainforest are used to pack dried fish and other market wares.

The new settlers, many of whom come from the overpopulated Sahel region, have no cultural ties to the forest and are not capable of dealing with its sensitive soil.

Old Knowledge – New Science

For some years now, worldwide efforts have been made to document the traditional knowledge held by many forest-dwelling people. In the face of forest destruction around the globe, botanists and pharmacologists have begun to search for new organic substances before they are lost. The traditional knowledge of forest people is an invaluable aid and in the course of their work, it has become evident that the world's rainforests hold an incredible number of useful substances. A worldwide search is underway for plants containing substances capable of fighting cancer. A number of African plants may also produce new medicines. The economic value of new plant medicines should not be underestimated. Some years ago, more than a quarter of the 1.5 billion prescriptions written annually in the United States were already based on substances gained from medicinal plants [116].

Ethnobotany, the study of traditional knowledge concerning plants and their use by native peoples, has become a popular branch of science. The majority of ethnobotanical studies have been made among Mexican Indians and in Amazonia. A few studies have also been made in Asia. But the new science has apparently not yet spread to Africa. A closer look, however, shows that some botanists had already done some very thorough ethnobotanical work in the years preceding World War II. J. M. Dalziel, a doctor and botanist who worked for many years in West African colonies, identified more than 900 genera of useful plants in West Africa, many of which included several species. His work of over 600 pages was based on notes made of West African traditions and included many of the ways in which the plants were used then and sometimes still today. Some of his sources dated back to 1905. But his information did not refer only to the rainforest, traditions in savannah areas were also found to be documented [94]. In addition, F. R. Irvine published surprisingly detailed information on the traditional uses of woody plants in Ghana [44]. Irvine worked at the Herbarium of the Achimota College in Ghana before World War II and later at the University of Edinburgh. According to his information, the leaves of 106 and the fruits of 326 woody plants were included in the local diet. Oils, fats and waxes were extracted from 84 species. Parts of no less than 755 woody plants were used for medicinal purposes. The majority of the species listed are a part of rainforest flora.

A Pharmacy in the Forest

Today, medicinal plants continue to play an enormous role in Africa. Local health care in many countries is still practiced mostly by traditional healers, medicine men and "barefoot doctors". Washes and pastes made from leaves, bark and roots are commonly used to treat wounds almost everywhere in Africa. But the treatment by natural remedies in Africa does not get the attention it deserves. Traditional healing is often accompanied by witchcraft. Medicine men also perform ritual ceremonies and recite magical incantations. Not only do they treat earthly ills, but they are also called upon to exorcise evil spirits. This obscure combination of magic and natural medicine has met with the scepticism of western civilization. And it is for this reason that western medicine considers traditional African medicine no more effective than a fairy tale when it comes to healing. Herbalists and medicine men are labeled charlatans. But such arrogance can have serious consequences

Not every stick is suitable for pounding Fufu, a common dish in West Africa. The slender trunks of young *Celtis mildbraedii* are popular for the task. They are numerous enough in Nkwanta, a forest village in southwestern Ghana.

for many developing countries. The low status of natural medicine has hindered traditional knowledge from being passed on as widely as in the past.

Africa suffers daily economic and cultural losses through the death of medicine men who have not passed on their secrets. The loss of traditional knowledge concerning the uses and effects of medicinal plants has been compared with the burning of entire medical libraries. Natural medicine is not only less expensive, it also includes substances effective against illnesses for which western medicine has not yet found a cure. One example are the leaves of the spreading shrub *Combretum mucronatum*; they can be used to expel the Guinea worm, a much-feared parasitic threadworm which embeds itself and eats away at skin tissue. Clinical tests at the Center for Scientific Plant Medicine in Ghana have only recently confirmed what traditional healers have

long known. The effects of traditional medicinal plants against diabetes and bronchial asthma have also been confirmed [116].

African governments would at least be well advised to examine the medicinal plants used in their countries. Nationally, they could help save costs on health care. But there is also growing international interest in medicinal plants and the World Health Organization (WHO) has begun to study the topic. In Cameroon, the Centre for the Study of Medicinal Plants (CEPM) has started an inventory of the medicinal plants used in the country and is establishing a herbarium. Medical fluids, tablets and ointments are already being produced in larger quantities for commercial purposes; one example is an anti-bacterial ointment which contains an extract from a species of *Erythrina* [117].

Professor Edward Ayensu, former Director of the Species Conservation Program at the Smithson-

Fuelwood is a critical problem in dry tropical zones. Rainforests provide a nearly inexhaustible supply of fuelwood.

ian Institution in Washington, D. C., has examined the most commonly used medicinal plants in West Africa. A native Ghanaian, Ayensu is equally interested in both rainforest conservation and the link between traditional and modern medicine [116]: He lists 187 medicinal plants used in one way or the other against more than 300 different pathological symptoms. Some of the plants are very popular and have a variety of uses. The Ashanti pepper (*Piper guineensis*) is quite a versatile plant. Soup made from its leaves is supposed to help women to conceive, soaked leaves are effective against coughs and used as warm compresses on wounds. Pulverized twigs and bark are used against coughs, bronchitis, – and as an enema against intestinal infections. The root relieves the pains of gonorrhea and is also effective against bronchitis. The small brownish-red fruit is similar to black pepper and used as a spice which is sold at local markets as "Bush Pepper". But aside from its culinary aspect, the fruit is supposedly effective against tumors and rheumatism. And finally, the pulverized seeds of the small fruit help relieve back pains, syphilis and rid clothes of insect pests [116,94]. The healing properties of Ashanti pepper may be overestimated, but there is surely some truth to the mass of information passed on over the years concerning this humble climber found on the trunks of trees deep inside closed rainforests.

Trade in Secondary Forest Products

The inhabitants of West African forests were not the only ones interested in products gathered from the forest. During the last century, products from West African rainforests began to attract much larger markets in Europe and North America. Long before rising industrialized countries discovered the forest's vital substance – tropical timber, major trade had developed with the West African coast in spite of strong fluctuations due to political upheavals and the price offered for the commodities in Europe. The major export products gathered in the rainforest were gum copal, rubber and cola nuts.

Gum Copal

"Copal hunters" collected the gum in the wild from trees of the genus *Daniellia*. Lumps of gum copal were found buried at the base of the trunks and on wounded trees. In Ghana, copal production was concentrated in the forests surrounding Akropong where a branch of the Basle

Market day in Asawinso in the rainforests of western Ghana. A medicine man recommends essences, bits of fur, nuts, horn, bones and the feathers of certain forest animals. Here, one openly admits that medicine and magic go hand in hand.

Mission was located. The Basle missionaries were known for developing African trade with Europe. In 1850, gum copal was first exported to England for the production of varnish, veneer, paint and linoleum. The largest export volume of about 500 tons was reached in 1876, a quantity never again surpassed. Price fluctuations on the European market caused hardship for the exporters. Towards the end of the 1880s and just before World War I, the copal hunters again enjoyed an increased demand, this time from the United States as well as from Europe. In the following years, however, production sank until it eventually disappeared entirely in 1936 [1]. From 1920 onwards, Congo copal, found in the swamp forests along the courses of the Congo tributaries, had begun to vie with the West African gum-resin. Congo copal can be found buried up to one meter below ground at the base of *Copaifera demeusei* [118]. But today, artificial resins have also replaced the demand for this gum copal.

Rubber

The West African rubber tree *Funtumia elastica* and the liana *Landolphia owariensis* exude a milky fluid (latex) which immediately hardens to a firm, rubbery mass. This raw rubber is of comparable quality with *Hevea* rubber, tapped from a Brazilian tree (*Hevea brasiliensis*) widely cultivated throughout the tropical world today. The significance of West African rubber was first recognized in 1883 and the best quality was found in the Krepi area of southern Ghana. Rubber was exported from Accra under the term "Accra Biscuits" mostly for tire production. But rubber tappers did not always treat the *Funtumia* trees with care. Instead of being tapped, the trees were often felled. The collectors then built a fire beneath one end of the trunk to increase the sap flow from the other end, a destructive practice which was forbidden by some village

chiefs on their tribal land. Once the Ashantis became active in the rubber industry, Ghana rose to become the third most important rubber producer worldwide. Rubber was partly produced in plantations of West African rubber trees. After a last increased demand from England during World War I, production sank, however, and finally collapsed or made way for *Hevea* rubber plantations [1].

Cola Nuts

In Ghana, cola nuts did not become a popular export product until this century when their export volume surpassed that of all other forest products gathered from the wild. 13 000 tons of cola nuts were exported in 1921. They were imported mainly by the Maghrib countries of North Africa. Cola nuts had been transported to the north in much earlier times over the Sahara desert. The highest prices were paid for nuts of *Cola nitida*, which have a stimulating effect when chewed. Since 1870, the major market for cola nuts has been Lagos which created such a steady demand that cola plantations were established in some areas, notably in Ho and Kpandu in Ghana [1].

The Oil Palm Conquers the World

Gum copal, *Funtumia* rubber and the cola nut are exports of the past. These products which can be extracted with little or no damage to the forest are no longer valuable on the export market. They have made way for other commercial products, the production of which causes much greater damage to the forest: cocoa, coffee, *Hevea* rubber and tropical timber extraction. Large areas of forest were also sacrificed for plantations of oil palm (*Elaeis guineensis*), a tree native to West Africa. Earlier, the fruit of the oil palm was gathered in the wild. The first plantations were established in Ghana around 1850

Oil palm fruits being delivered to the oil mill on the PAMOL plantation near Mundemba (Cameroon).

and led to a somewhat more steady export revenue. Palm oil is pressed from the red, fibrous pulp (pericarp) of the fruit and palm kernel oil is extracted from the oil-rich seeds. The export of both products provided foreign exchange income to West Africa. Today, 14% of all plant oils worldwide is derived from the oil palm, which is nearly equal to the proportion covered by the soya bean and comparable to that of the sunflower. But since the oil palm grows throughout the year, the yield per hectare is much greater, from 1000 to 4000 kg and sometimes up to 6000 kg [119]. Malaysia and Indonesia are responsible for more than half the current world production of palm oil. Other producers, including West African countries, suffer from the competition. The situation was different not too long ago. Although the oil palm was introduced to Malaysia 75 years ago, production was not satisfactory. The trees needed to be pollinated man-

ually, which was a tedious process and led to inefficient production. It was originally believed that oil palms were wind-pollinated until Dr. R. A. Syed discovered the contrary. The Commonwealth Institute for Biological Control assigned Syed to study the pollination of oil palms in a PAMOL (Unilever) plantation in Cameroon. He found that the pollination was effected by insects and not by wind! The weevil-beetle *Elaeidobius kamerunicus* in particular plays an important role in the pollination process. After cautious tests, the release of the beetle in the Malaysian plantations increased the yield by 40–60%. The West African weevil-beetle proved to be a gold-bug for Malaysia, it raised the foreign exchange income by US$ 44 million in the first year [119]. Today, biotechnological methods permit the cloning of oil palm tissue. Cloning allows the vegetative multiplication of the germs from especially productive plants. Thanks to biotechnology, production has sky-rocketed while prices have dropped – not to the advantage of the oil palm's native home: West African producers are losing out. The PAMOL plantation in Cameroon, on which the valuable weevil was discovered, went bankrupt in 1987.

West Africa has given much to the world and received little in return. But there are products yet to be discovered in the rainforest which may someday be of commercial value. A shrub distributed from Ghana through to Central Africa bears an incredibly sweet fruit: the "miraculous berry" (*Synsepalum dulcificum*). In the Volta Region of Ghana, the local people eat this fruit which initially has a sweet-sour taste. Surprisingly, the sweetness of the miraculous berry is retained in the mouth so long that even substances as bitter as quinine can be camouflaged up to an hour later. The miraculous berry is also used to sweeten palm wine [44]. The British company Tate and Lyle is also interested in the super-sweetening agent. The sweetness is apparently not due to sugar content but to a protein complex [119]. Nevertheless, one cannot assume that the commercial value of the miraculous berry or of any other secondary forest product would be reason enough to permanently protect the forest where these plants grow. West African rainforests have been and continue to be used commercially without thought to their future or to that of the people who depend on the forest for their livelihood.

The Economic Value of Traditional Hunting

On dark, moonless nights West African hunters leave their villages and venture into the forest despite their fear. Moonlight filtering through to the forest floor would hinder the hunt. The hunter sees his prey by its reflecting eyes in the light of his carbide headlamp. The blinded animal is not to see that a man stands behind the glaring light. Most of the nightly catch consists of duiker antelopes, usually the small blue duiker which seldoms weighs more than 10 kg. Not that hunters have a specific preference for the smallest of the duikers, they simply can never be sure of their prey until it lies at their feet in the light of the carbide lamp. In many areas, the blue duiker is the most common hoofed animal. Occasionally, a hunter may bag a bongo, the largest of the forest antelopes. This animal with its attractive coloring and massive spiral horns is usually very cautious and seldom seen by man. It is rarely listed under the catch of West African hunters. But should a hunter slay such a majestic animal on a dark night, he will think twice of how to deal with the unusual situation. In Ghana, forest-dwelling people consider the bongo a sacred animal and the successful hunter will fear for his life. He may not bring his prey into the village at dawn. He must skin it outside the village and under no circumstances should he carry the head of the bongo on his

own head. Afterwards, the hunter must bath in "Sassandra", a spiritual medicine to prevent him from either going mad or dying. Spirits may accompany him on his next hunt. The Gouro tribe, who live along the middle course of the Bandama River in the Côte d'Ivoire, believe in the reincarnation of all men and animals. The same souls are born again and again. Whoever kills a man or an animal must reckon with the soul's revenge depending upon the strength of its "Bei". Bei, which more or less means "power", is not of equal strength in all animals. The leopard and the elephant have the most dangerous Bei.

The night hunter with his carbide headlamp has learned to live with fear. Many a hunter has lost his life in a sudden encounter with a forest elephant. The elephant, blinded by the glaring light, allows the unwitting hunter to approach and then attacks in fright, suddenly surprised by the man's proximity.

In order to prevent their Bei from attacking him while he sleeps, a hunter must make a sacrifice to the head of the slain animal immediately after the killing. The skulls are then piled at the base of a fetish tree outside the village [115].

Yoruba hunters also fear the revenge of an animal's soul. The Yorubas are native to southwestern Nigeria. When a hunter kills a leopard, he binds its eyes because the animal "is a king and his gaze is too frightful for the people" [115]. It is, however, unlikely that there are many leopards left in the Yorubas' native territory today. The rainforest has given way to an increased human population density. But other West African tribes also believe in the special powers of the leopard. The fur of the animal must be presented to the tribal or village chief.

Hunting Taboos and Varied Prey

It is often forbidden to hunt or eat specific animals in West Africa. Such taboos usually pertain to individuals or families. As its name implies, only a chief may eat the meat of the royal antelope, Africa's tiniest ungulate. Hunting chimpanzees is often totally banned because of its similarity to man and the hunters' fear of its soul. A village chief among the Sefwis in western Ghana admits, however, he would welcome a hunter's bringing him the skin of a chimpanzee for the great village drums. Drum skins from chimpanzees are said to be particularly resilient and can be beaten for four years before having to be replaced. Along the West African coast and rivers, the python is often considered sacred. It symbolizes the spirits of the water and of war, fertility and wisdom. In today's Benin, it was once custom for traditional African priests to keep pythons [52].

There are few hunting taboos, however, which are exclusive and respected by all rainforest people or even by an entire tribe in West Africa. When the prey is worth more than the cartridge and the hunter is able to overcome his fear, practically no animal of the forest is spared. Monkeys active during the daylight hours are hunted at day. In the Korup area of West Cameroon, as in other areas where the red colobus has not yet been wiped out, this monkey species is part of the most common catch. Drills (Papio leucophaeus) are also found in this scarcely accessible area of rainforest. The large groups of these ground-dwelling pavians are hunted with the aid of dogs. In order to prevent the powerful pavians from grabbing the dogs by the tail and killing them, the hunters dock the tails of their hunting companions. The dogs cut pavians off from their group and chase them up trees, where they are sitting prey for the hunters [120].

Wire snares are sometimes used to catch rodents, monkeys and duiker antelopes. Reptiles are hunted with bush knives. The diets of some tribes in Ghana include squirrels, flying squirrels, pangolins and even fruit bats. The large maggot of the palm beetle (Phyncophorus sp.) is also considered a delicacy by some forest people [121]. Crocodiles, monitor lizards, snakes and the African giant snail in particular are also popular sources of meat. Europeans tend to poke fun at the omnivorous habits of many West Africans. A decent person supposedly eats beef and poultry. Consequently, the value of the rainforest as a source of protein is often greatly misjudged and underestimated. It is interesting, however, that bushmeat is in much higher demand and more expensive than domestic meat in most of West Africa. Even much of the urban population still prefers wild game.

Bushmeat – An Invaluable Resource

"One of the ironies of the age is the tendency of scientifically educated people to ignore or reject whatever they cannot measure. Such is the case with the consumption of wildlife of all kinds for

Western medicine lies on the desk in front of the school-master of Fabe, a settlement in southwestern Cameroon. The skull of a drill on the wall above is a symbol of the people's dependency on the forest and the picture of the president reminds of civil duties – contrasting ways of life in need of daily integration.

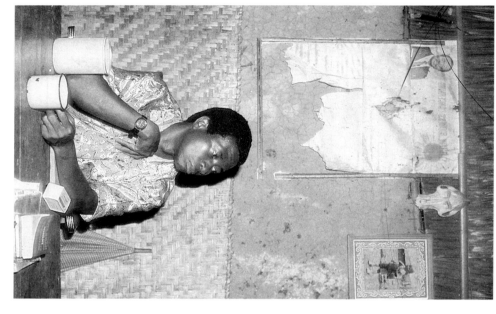

ideas and things which are often foreign, distant and unconnected to the lives of those whom they want to help. "These words were written by the editor of "Unasylva", a periodical published by the Food and Agriculture Organization (FAO) to introduce a series of articles dealing with the utilization of game as a source of food. The editor was apparently critical of his own organization. A survey of nutritional habits showed that in Africa, and particularly West Africa, a surprising variety of game continues to be a major source of protein [122]. The FAO normally deals more with aspects of timber exploitation in tropical rainforests, a practice which in turn often considerably reduces the abundance of wildlife.

Sunday S. Ajayi, from the University of Ibadan in Nigeria, and Emmanuel E. O. Asibey, former director of the Ghana Forestry Commission, have both statistically proven that game (or "bushmeat" as it is called in English-speaking parts of West Africa) is actually a very important factor for local economies. Game consumed in Nigeria during the early 1970s was worth £30 million and corresponded to 4% of the country's gross national product [123]. In 1980, estimates of Nigerian trade in game fluctuated between 150–3600 million Naira (1 Naira = US$ 1.00). The range of these estimates [124] mirrors the difficulty in judging a market which does not officially exist. Although the estimates point to several percent of the country's GNP and 95% of the population surveyed reported that they regularly or occasionally consume bushmeat, game scarcely appears in the statistics. Since it does not fall into a trade category, it is not considered a forest product of economical value. It is a non-product as are other "secondary forest products" as well. Emmanuel Asibey complains, "Unfortunately, the Africans themselves tend to fail to appreciate, and insist on, the inclusion of bushmeat production in development plans" [125]. He has long pointed out the

food in developing countries. The significance of this resource is largely ignored by nutritionists, animal production experts and even some wildlife biologists because it is difficult to find statistics about the gathering, marketing and consumption of wildlife. In addition, these foods are mostly strange or even repugnant to the majority of specialists who are working so hard to increase food production and human nutrition levels among the peoples of developing countries. The specialists are inclined to think of the improvement of man's lot in terms of passing on what they are familiar with in their own lives,

Wire snares (photo left) are most often used to trap duiker antelopes, brush-tailed porcupines and giant rats (photo right).

extreme importance of this kind of land-use. In Ghana, 75% of the population relies on traditional sources of protein: Game, fish, insects, maggots and snails [126].

Game is popular throughout West Africa and is not considered a substitute for better quality meat. In Nigeria, for example, bushmeat from indigenous forests commands a higher price than does any domestically produced meat. Only the best cuts of imported steak cost more than bushmeat in a Nigerian supermarket. Even the wealthier urban population prefers the meat of a cane rat or duiker to that of a goat or sheep.

Urban eating habits still reveal the rural origins of the city's inhabitants, much more than their other behavior.

But the urban supply of game meat is not always sufficient. Nigeria's upper class does not consume as much bushmeat as does the rural population although the wealthier members of society would gladly pay the price. The majority of game meat is traded and consumed locally. Many hunters and trappers take their catch to the edge of the next cross-country road and hold it aloft. They are fairly certain to be relieved of their burden by one of the next passing vehicles. A survey in Bendel State in southern Nigeria disclosed the variety of the roadside market: Monkeys, genet cats, mongooses, bush-pigs, pangolins, flying squirrels, birds, tortoises and even snakes. Three quarters of the catch, however, consists of duiker antelopes, giant rats, brush-tailed porcupines and the much sought after cane rats [124]. The abundance of the latter is due to the opening up of the forest and the increase of grassy areas. In Ghana, too, the large cane rat, locally known as "grasscutter", is the most popular form of bushmeat. At the Kantamanto market, one of the largest markets in the capital of Accra, some 24000 of these good-sized rodents were sold from December 1968 to June 1970 which corresponded to a live weight of 117 tons and a market value of US$ 125000 [126]. These figures led Emmanuel Asibey to test the domestic breeding of large cane rats as a source of meat in the 1970s. His experiments were quite successful and showed that cane rat breeding could supply extra income to farmers. The rodents are easy to keep and require only grass fodder [127]. Unfortunately, large-scale cane rat breeding has never become popular although it would be an efficient method of meat production, especially over wide expanses of West Africa where grassy areas are scattered throughout fallow forest.

Game Guarantees Local Income

Contrary to the prestigious sport hunter of the temperate zones, a West African hunter does not boast of his catch. For him, hunting is a way of life and not a pastime for discussion. Neither does he dress for the occasion with a shooting jacket but wears the least valuable clothes he owns. Although hunting guarantees a respectable income, a hunter is not necessarily a respected man. Young people in the village consider the job dangerous, difficult, dirty and old-fashioned. Few strive to become hunters [120].

At most, a hunter is admired for his weapon but not for his occupation. Fear of the animals and the invisible dwarfs, who can never be trusted, keep hunters modest and reluctant to speak of their actions. Many traditional hunters' fear of breaking the law, knowingly or not, adds to their uneasiness. Indeed, many West African nations limit hunting to specific species, seasons and hunting methods. The use of arms usually requires a permit. And in some areas such as national parks, hunting is prohibited entirely. It is understandable that West African hunters remain silent when asked about their work. Most earn their living as farmers and stalk animals only to supply their families with meat or to augment their agricultural income. It takes time to gain the trust of a hunter before he will tell of the tradition and his hunting habits.

At the suggestion of Emmanuel Asibey, the trust of several traditional hunters was slowly gained in order to study the importance of game-hunting for local economies. A considerable part of the population in Kwamebikrom, a small village north of Bia National Park in western Ghana, lives wholly or in part from hunting. Allowing for uncertainties, one can assume that 50 wild animals are killed each month by the people of Kwamebikrom. In 1978, the resulting market value corresponded to 8 or 9 minimal monthly salaries (Tab. 13). The majority of the meat is sold fresh or smoked to traders. The people cover their own need for protein with other sources not included in the study: Fish, smaller animals caught in wire snares and especially giant snails. In relatively undisturbed forest areas, game thus probably provides 20–50% of the village cash income in addition to covering private nutritional needs.

A close examination of hunting habits in the Korup area of West Cameroon showed that approximately 38% of a village's income is provided by hunting. Trapping provides an additional 18% of the income. Hundreds to thousands

West African hunters. They have little in common with the sportly dressed huntsmen we know.

Table 13
Game caught by inhabitants of the forest village Kwamebi-krom (Ghana) from May to August 1978 [66]. The legal minimum wage at the time amounted to 112 Cedis/month.

	Number of Animals	Local Market Value (Cedis)
Blue duikers	50	1875.—
Campbell's mona monkeys	46	464.60
Royal antelopes	33	389.40
Lesser spot-nosed monkeys	11	115.60
Bay duikers	10	425.—
White-crowned mangabeys	9	135.—
Brush-tailed porcupines	9	108.—
Giant rats	3	15.—
Olive colobus	3	30.—
Diana monkeys	2	30.—
Black-and-white colobus	2	24.—
African civets	2	80.—
Flying squirrels	2	14.—
West African dwarf crocodiles	2	40.—
Nile monitor lizard	1	20.—
Pangolin	1	10.—
Bush-pig	1	50.—
Large cane rat	1	12.—
Giant forest squirrel	1	9.—
Total	189	3846.60

Table 14
Game prepared and served during 1976 in the "As Usual" chop bar in Sunyani (Ghana) [120].

	Number of Animals
Large cane rats	1484
Blue duikers	528
Bushbucks	316
Brush-tailed porcupines	167
Giant rats	126
Royal antelopes	122
Flying squirrels	96
Bay duikers	92
Black duikers	42
Two-spotted palm civets	36
Lesser spot-nosed monkeys	34
Red-flanked duikers	31
African civets	30
Pangolins	15
Francolins	13
Guinea-fowl	11
Campbell's mona monkeys	11
Genet cats	9
Black-and-white colobus	9
Total	3172

of traps are laid especially during the rainy seasons. The combined annual catch from hunting and trapping amounts to at least 217 kg of meat per square kilometer [120].

Local restaurants in larger towns and in West African cities serve large quantities of bushmeat: In Ghana, such restaurants are called "chop bars". "Fufu", a paste-like mass made by pounding cooked cassava, coco-yams and plantains, is served as an accompaniment to generous amounts of bushmeat in a hot, spicy soup. At the chop bar "As Usual" in Sunyani, over 3000 wild animals were consumed during one year (1976). Since Sunyani lies in the transitional zone between rainforest and savannah, the menu at "As Usual" often lists savannah animals: bushbucks, Guinea-fowl and especially the popular

large cane rats (Tab. 14). A total of 123295 meals were prepared with 14630 kg of game meat and were sold for almost 78000 Cedis in the chop bar alone. (78000 Cedis then corresponded to about US$ 68500.) About two-thirds of the cash income went directly to the full or part-time hunters who sold their catch to the chop bar [121]. Close to 80 farmers in Sunyani earned as much money this way as does a government worker. For the farmers, however, this is a supplement to their income from subsistence and cash agriculture. According to Asibey, most small-scale farmers would not be able to continue cocoa production if it were not for the extra income. Since 90% of Ghanaian cocoa is grown on small-scale farms, the harvest would fall far below the present level without the additional income from wildlife utilization. Thus, the rainforest's wild animals, which are usually overlooked in land-use planning or at best classified as secondary forest products, are not of such secondary value. Not only is game meat the most important source of protein and of local economic value which should not be underestimated – indirectly, game even subsidizes the export economy.

The Great Demand for Giant Snails

African giant snails are a kind of bushmeat which is easily gathered. Large quantities of snail meat are eaten in West Africa and it is incredibly popular with forest inhabitants in Nigeria, Ghana and the Côte d'Ivoire. Plantation workers bake them in the shell over a campfire. Shelled, the snails can be roasted or cooked. Snail meat is often skewered and smoked to preserve for longer periods before it is prepared with okra (lady finger) or another vegetable and served with fufu. At first, the meat may seem somewhat rubbery, but one soon learns to appreciate its flavor. Snail meat not only tastes good, its protein content is comparable to that of beef. It is even richer than chicken eggs in certain essential aminoacids, notably arginine and lysine [128].

Snails do not only feed the poor; even in Abidjan, in many respects West Africa's most modern city, rainforest snails are eaten in large quantities. The city likens a French metropolis with its silhouette of sky-scrapers, the department store chain "Uniprix" and an artificial ice rink located atop the luxury hotel "Ivoire". Abidjan is the major trade center of the Côte d'Ivoire and provides a home to about 20% of the nation's 10 million citizens. Eating habits, however, have remained traditionally rural. What may seem to appear a provocation in one of the city's elegant French restaurants is all the more appreciated at home. A survey by the Central Laboratory for

Animal Production LACENA in Abidjan showed that in the city alone, one million kilograms of giant snails are eaten annually [129]. The majority is consumed by people belonging to the forest tribes of the Baoulé, Agni, Bété, Krou, Bakwé, Dida, Guéré, Yakouba and Gouro. Although urbanized, these people have retained their preferences for forest delicacies. During the dry season when fresh snails are a rare commodity, these customers are prepared to pay more than double the price for beef. But even during the wetter months, snails are never cheaper than beef at Abidjan's eight marketplaces.

Approximately half of the giant snails consumed in Abidjan originate in the forests around Céchi, 140 kilometers inland. They are gathered and packed live into bags which are then transported by truck to the capital. But snails are also delivered from all parts of the country's forest area to Abidjan. All over the rainforest, snail-gathering is a traditional pastime. Young and old take part in the gathering during the farming periods when families leave their villages to temporarily live in farming camps. Snails collected in the forest surrounding plantations and not destined for private use are stored in the camps and head-carried to the weekly market.

Snails are abundant in relatively undisturbed forest areas where entire familes can earn a living by gathering this sustainable product. Besides the gatherers, wholesale traders and retailers also profit from the snail business. Women are usually responsible for local trade and sale all over West Africa. In the Côte d'Ivoire, the total weight of snails sold in 1986 was estimated at 7.9 million kg and thus accounted for 10% of the trade in game meat [129]. The proportion of snail meat in relation to the total game consumed in many regions of West Africa may be even higher.

African Giant Snails – Domestic Potential?

The largest land snails belong to the family of the African giant snails (Achatinidae). In recent years, the common giant snail (Achatina fulica) has made headlines in Asia and Florida where it was unwittingly introduced and has since become a nearly uncontrollable pest. Various species of giant snail occur in West African rainforests: Achatina achatina, the "true" giant snail, and Archachatina ventricosa are the most important for human nutritional needs. These two species occur only in intact rainforests where they depend on the relatively high humidity, the constant temperature of around 26°C and an absence of wind. There are strong local and seasonal fluctuations in their numbers. During the main rainy season in May, snail densities may reach 2.2 individuals per square meter. These rainforest creatures mainly feed on decaying leaves and the fruits of various trees. Most giant snails hibernate during the dry season. Giant snails grow and reproduce rapidly. Breeding experiments with Achatina achatina showed an average reproduction rate of 26 offspring annually. But keeping and breeding giant snails is not easy. Snail eggs require an extremely constant incubation environment. The production of giant snails as a source of protein is liable to continue to depend on intact rainforests [128].

Since empty snailshells are too unpleasant in smell at first, they are not tolerated near the village but are piled along a logging road.

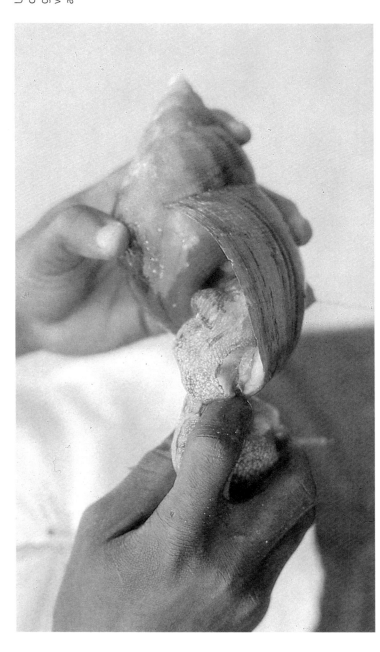

Unworthy of a listing in official statistics – the African giant snail. But it is of great value for human nutrition and the local economy.

Misunderstanding Forest Products

The fact that wildlife utilization and forest product gathering are still rated as being of minor importance is related to the unilateral export orientation of national economies. West African countries have remained export oriented and thus dependent on Europe as in past days of colonialism. Today, however, the income from timber export is dropping due to reduced supplies. Secondary forest products must all the more be given sufficient consideration in future land-use planning. "Team approach towards wildlife and forest management has to be accepted at the highest forestry decision-making levels and filtered through to the grassroots of both disciplines, particularly forestry whose staff have always looked to wildlife in their management unit as a fringe benefit of the service". This claim raised by Emmanuel Asibey of

Ghana's Forestry Commission, should be heeded [121]. Further studies on the value of game and secondary forests products will be necessary.

In order to guarantee a sustainable use of wildlife resources, the effects of hunting on the animal populations must also be examined. Although the potential for production of game meat from West African rainforests has been grossly underestimated, certain species react vulnerably and may be wiped out if hunting is not controlled. Top canopy monkey species, the red colobus in particular, have already disappeared from much of West Africa's forest. If game populations are to be preserved, hunting and trapping doubtlessly need to be limited within most of Africa's remaining rainforests. Various recommendations have been made [130]. The allocation of quotas for certain forest areas, the limitation of hunting licenses

A Liberian game guard has confiscated two legs of a banded duiker, a protected species. Confiscated or not, they are sure to land in a cooking pot.

and even a total kill to be determined on the basis of wildlife population estimates. But how are hunting regulations to be enforced in a rainforest where the range of vision measures at most 15 meters and where the human population lives to a good deal from bushmeat? There are already a number of hunting regulations in West Africa: In Ghana, it is legally prohibited to hunt certain species and a general hunting ban extends from August to December. The ban goes unheeded – who is prepared to refrain from eating meat four months of the year? Or do the legislators expect the people to slaughter the country's entire population of chickens, goats and sheep during those four months? Unthinking conservationists even induced Liberia's head of state to declare a total hunting ban in 1988. It was soon revoked, however, in favor of a shorter list of protected species.

Similar to the rest of West Africa, traditional hunting and trapping are of great importance in Cameroon. Hunting by means of traditional methods excluding firearms does not require a license. This includes spears, bows and arrows, as well as traps fabricated from local materials. The pygmies, who live largely from hunting, do not require permits of any kind [131]. But despite the liberal nature of Cameroon's hunting regulations, village income in West Cameroon would greatly suffer if they were strictly enforced [120].

"A law is only as good as its enforcement." The truth to this statement can be found everywhere in West Africa, but it is of special importance in the rainforest. It would be an illusion to believe in the enforcement of laws imported from abroad and not in accordance with tradition. The best wildlife protection lies within the rainforest itself. If past decision-makers had sufficiently concerned themselves with forest protection in decades gone by, hunting regulations would hardly be necessary today.

It is too bad for the Malimbe bird, but whoever wants to become a West African hunter first practices with a slingshot.

Sacrificing Forest for Short-term Gain

∨ The story of West Africa's tropical timber industry began with the reddish wood of the African mahogany tree. Originally, these trees were felled only in the immediate vicinity of large rivers so that the logs could be rafted to the coast. Large-scale exploitation was not possible until road transportation became a viable alternative.

In the first half of the 1980s, an annual forest loss of 7200 square kilometers was recorded in the countries along the Gulf of Guinea, a figure which corresponded to 4–5% of the total remaining rainforest area. In 1985, 72% of West Africa's rainforests had been transformed into fallow territory and an additional 9% had been opened up by timber exploitation (Tabs. 1 & 2). The future of West African rainforests as a living form of vegetation and a habitat for both man and animals is thus uncertain today. The biological diversity and ecological equilibrium of the entire region is endangered. Some scientists fear that regional climate change has already begun due to large-scale deforestation and that what is left will be jeopardized by longer dry periods in the future. Foresters and the timber industry are discontent over forest destruction and an increasingly short supply of valuable hardwoods while the domestic timber needs of these countries are rising. Have we entered the last stage of a 500-year exploitation of the rainforest along the Gulf of Guinea?

Forest loss is often attributed to increasing human population and, indeed, the influx of farmers from the north worsens the situation drastically. In the Côte d'Ivoire, for example, hundreds of thousands of farmers from Mali and Burkina Faso emigrated between 1966 and 1980 to the rapidly expanding areas of timber

Immigrant farmers follow in the wake of timber exploitation. They have no feeling for the forest and often clear large tracts of land like this one in the northern part of Ghana's rainforest zone.

< The meager soil is fertilized by the ash resulting from slash-and-burn clearing – but not for long.

exploitation. The area of forest damaged by slash-and-burn agriculture almost doubled and affected 90% of the entire rainforest zone in 1980 [8]. In the southwestern Côte d'Ivoire, where Taï National Park was established in 1972, the human population density had been recorded at an average of 1.3 persons per square kilometer. The inhabitants belonged to the tribes Krou, Bakwé and Bété. Eight years later, the population density had risen to 7.7 persons per square kilometer [132]. Timber exploitation had opened up the region and large-scale agricultural projects had attracted farmers from the north thereby making the original forest inhabitants a minority. The completion of the Buyo dam on the Sassandra River led additional settlers from the flooded regions to move into the area surrounding Taï National Park.

Overpopulation in the Sahel countries was not the only factor leading to the invasion of new settlers, however. The countries along the coast were themselves partly responsible. From 1969 to 1970, Ghana deported 300 000 Nigerians and citizens of other West African countries. Fifteen years later, Nigeria revenged itself by employing the same tactic and deporting thousands of teachers and trained workers of Ghanaian nationality. Ghana's economic crisis of the early 1980s left many with no choice but to look to the forest for a new way of life.

Immigrant farmers clear the forest bit by bit using axes and bush knives, leaving the dried remains of cut vegetation to be burned. The immigrants from Sahel countries are especially unfamiliar with the forest's sensitive soil. After only a few years harvest, it becomes depleted and infertile. Consequently, the farmer follows the path left by timber exploiters and moves deeper into the forest to clear his next plots of

Farmers mark their temporary activity at the edge of a timber transport road. First, the Europeans brought Christianity, then the roads.

A mixed plot of cassava (slender, knotted stems left), plantains and coco-yams (*Xanthosoma sagittifolium*). Both the tubular roots and the large, heart-shaped leaves of the coco-yam plant are used to prepare tasty meals.

land. The exhaustive use of land for the temporary production of corn, plantains, cassava, yams and a few vegetables leads to an ever-increasing area of fallow land. A few tall trees and groups of smaller trees are all that is left protruding above high grass and thick, scrubby brush. A chaotic mosaic of farm plots fills out the picture. Badly scared wooded hills, eroded slopes and the ruins of mud wall huts tell of the abandoned crop beds. Such land must lie fallow for some years before it can be cleared again and replanted. Even then, it will yield but a few modest harvests. If the land is not too sloped and has not been entirely eroded of its soil, it will only regenerate after a period of many long years.

The native gatherers and planters who had practiced shifting agriculture for centuries in the forest did not do irreparable damage to the land. They never cleared sloped ground and they worked only small areas. Afterwards, the plots were left to regenerate to full secondary forests.

There are such areas of secondary forest throughout West Africa and they are scarcely distinguishable in structure from closed primary forest. Some forest tribes in isolated areas have continued to practice shifting agriculture to this day and in over one hundred years, no lasting large-scale damage has been done to the forests surrounding their villages.

The slash-and-burn agriculture practiced by the immigrant farmers does not increase the area of cultivated land in proportion to forest loss. That is the irony of the rainforest's destruction. Since the soil can only be worked for a short time and cannot support sustainable use in most areas, the actual area of land cultivated is merely displaced as the forest retreats and does not increase to a meaningful extent. That is why there is so much fallow land, unable to support either timber production or agriculture. Neither are there forest products to be gathered, nor much game to be hunted. Thus, it is not a choice between forest or agriculture, as many may wish

to believe; rather, it is a choice between forest or a mosaic of farm plots with a great deal of fallow land inbetween.

Meanwhile, foresters and timber companies never cease to declare that they are responsible for removing comparatively few trees from the forest and that over 90% of the trees lost are sacrificied to slash-and-burn farming. Conservationists also sometimes divide the blame between the timber industry and the farmers according to who felled how many trees. Others mistakenly believe rainforest destruction to be connected with the problem of firewood. But

the latter is a difficulty in dry tropical zones, where the lack of firewood leads to the destruction of open forests and other dry woody vegetation. Collecting firewood is generally not a problem in the rainforest, even seriously disturbed forest provides sufficient dead wood.

It is not that easy to claim certain people or a single factor solely responsible for forest loss: Actually, the destruction began much earlier this century when the original forest inhabitants were stripped of their autonomy and the forest administration was centralized in the interest of commercial timber exploitation. Forest loss is the

A new timber road north of Sapo National Park, Liberia.

Table 15
Exploitation of undisturbed rainforests 1981-1985 [6].

	Undisturbed closed forests logged annually (ha)	Growing stock (volume of the bole of trees greater than 10 cm in diameter, m³/ha)	Volume actually commercialized (m³/ha)
West Africa	164 000	172	12
Central Africa	431 000	262	13

result of complex interactions between cultural and commercial factors which eventually lead to uncontrolled destruction. But the tropical timber industry plays a key role in the process.

The Consequences of Selective Exploitation

An undisturbed area of rainforest in West Africa harbors about 180 larger tree species (cf. Tab. 6). Fewer than one quarter of those species are of economic value and only 10 to 15 are actually marketable timber. In spite of efforts to encourage the use of lesser known timber species, the market continues to concentrate on a fraction of the usable timber available. In high-yield areas of exploitation in West Africa, there are usually more than 500 trees per hectare which measure a diameter of over 10 centimeters [28]. About one tenth of the trees are more than 30 centimeters thick and the largest specimens may have diameters of 200–300 centimeters above their buttresses. In spite of the fact that African rainforests have a higher growing stock than forests in other parts of the world, an average of only 12–13 cubic meters are harvested per hectare (Tab. 15). That is three times less than in Southeast Asia and does not even equal the trunk volume of one large tree. In forest reserves – most of which are open to exploitation – minimal diameters are prescribed for felling. In Ghana, that may be 68 or 120 centimeters depending on the species. But the

low area harvest is not due to regulations. It has more to do with the timber companies themselves, who are only interested in the most valuable species: They "cream" the forest. Extremely selective methods of timber exploitation lead to only 7% of the timber available in West African forests being felled and marketed.

One might think that this kind of hyper-selective exploitation would be the best guarantee for a sustainable timber industry. After all, the purpose of the selective system is the repeated harvest of the same area after a felling cycle of some decades! Reality, on the other hand, paints quite a different picture. Very selective exploitation is synonymous with very extensive exploitation and a low area yield quickens the pace at which further rainforest is opened up. Since the massive trunks can only be transported out of the forest over logging routes and feeder tracks, even selectively logged areas must be crisscrossed with a network of access roads. At least

First, the logger cuts away the lianas. They could become deadly whips when the tree falls. A wedge of the buttress is then cut away on the side of the assumed direction of fall (right side of photo). Then the logger begins to cut through all the buttresses surrounding the tree until finally, a cracking sound announces the giant's fall. The logger turns off the chain saw and loudly shouts a warning. The core of the trunk rips fiber by fiber, the giant crown above begins to sway and with a deafening roar, the many hundred years of a tree's history come to an end.

10 kilometers of road must be allowed for every 10 square kilometers of rainforest [4]. And they are the beginning of the end. Logging roads are the real reason why 90 % of the slash-and-burn activities by immigrant farmers is concentrated in exploited areas. Poorly staffed and underequipped forestry services are not able to supervise, let alone control, the fast pace at which forests are opened up and exploited, nor the influx of farmers. The forest guard is of little interest to the timber companies, at best they may give him a pitiful smile and perhaps a small bribe. Many firms themselves no longer believe that their logging concession will survive an entire felling cycle.

The Forestry Crisis

At the beginning of the century, foresters assumed that moist tropical forests could be managed in a sustainable fashion similar to Eu-

ropean forests. On the basis of this assumption, African forest administration was centralized and a system of forest reserves was established against the will of local inhabitants. Today at the end of the century, foresters are more perplexed than ever as to how rainforests should be sustainably utilized, if at all. They have themselves begun to doubt the sustainability of sylvicultural systems. Too much forest has been lost as a consequence of opening up forest for timber exploitation. The combined selective exploitation system became discredited in Ghana since many valuable secondary species were destroyed and large quantities of poison were used in the course of improvement thinning. From 1958 to 1970, 2590 square kilometers of rainforest were managed under this system during which time 188 tons of sodium arsenite were applied [133]. Other methods of improving timber yield were just as unsuccessful. After years of research and numerous tests, the Tropical Shelterwood System (TSS), which also required the use of sodium arsenite, proved to be a failure. Natural regeneration of valuable species under the TSS did not meet expectations. Enrichment Planting also proved unsuccessful. The seedlings of commercial species raised in tree nurseries at great expense and later planted in the forest were usually soon overgrown by ground herbs and vines. In both Nigeria and Ghana, the Tropical Shelterwood System and Enrichment Planting were abandoned in disappointment during the mid-1960s.

Tropical Foresters under Attack

Tropical foresters lead difficult lives today. They have been the object of international criticism and have been held responsible for rapid deforestation although they have always strived towards sustainable forest management. If forest-

ers can be accused of anything, it would be the fact that they were too concerned with the primary product timber and neglected the forest's multiple other functions. Although foresters are quick to point out the importance of secondary forest products, of game and gathered produce, those aspects were completely ignored in forest management practices. Officially, all of Ghana's commercially exploited forests in forest reserves were considered "managed". A study by FAO, however, showed that forest management in the Kakum Forest Reserve "has been geared towards timber production with scant attention given to the potential of minor forest products to stimulate rural industry and the use of branch wood and rejected logs for fuelwood and charcoal. Employment in forest industry is minimal and the concept of forestry for rural development has been ignored" [134]. These findings are in crass contrast to Kakum's reputation for model forest management.

Especially serious is the fact that tropical foresters too long deceived the world by claiming that moist tropical forests could be sustainably logged of a few valuable timber species. In the meantime, they were applying inefficient methods and thoughtlessly poisoning "worthless" trees and secondary timber species at the danger of wildlife populations.

Endangered Forest Reserves

The apparent failure of the foresters in their strive to exploit valuable species in a sustainable way led them to forsake the complex management of natural forests in favor of intensive plantation forestry. As early as the 1960s, large-scale plantations of native timber such as limba, emeri, obeche and opepe were established in Nigeria as well as of the exotic species gmelina, teak and pinus. The experience gained during the pre-war years proved useful in helping to set up the new plantations. According to the FAO, industrial timber plantations increased by approximately 71% in West Africa from 1981–85 (Tab. 16). Large-scale plantations were established especially in Nigeria and the Côte d'Ivoire. Unfortunately, this did not mean the remaining natural forests were spared. The Nigerians did

Table 16
Plantations of industrial timber in km² [6].
Firewood plantations, which were primarily established outside the rainforest zone, are not included.

	1980	1985 (estimated)
Benin	77.5	77.5
Ghana	262.5	319.5
Guinea	21.5	27.0
Guinea-Bissau	3.0	7.0
Côte d'Ivoire	378.0	658.0
Liberia	63.0	163.0
Nigeria	1463.0	2703.0
Sierra Leone	53.0	63.0
Togo	75.5	92.5
West Africa	2397.0	4110.5

◁ Besides African mahogany, the timber industry is mainly interested in the red *Entandrophragma* species sapele (SAP), gedu nohor (EDI) and utile (UTI). The stocks in West Africa are already largely exhausted.

Some Forest Management Systems Used in West Africa

Selection System

Under the selection system, timber trees of a minimal girth or diameter (example: 68 cm in Ghana) within a specific forest compartment are identified and marked on a stock map. Individual trees are selected for felling on the basis of the stock available and this is considered the yield of the specific area. Sometimes, non-commercial trees are also removed after exploitation, a practice meant to encourage the growth of valuable timber. The unwanted trees are frill-girdled. A strip of bark is cut from around the trunk and the tree is poisoned with sodium arsenite. The same area can be reharvested after a felling cycle of 15 to 40 years.

Tropical Shelterwood System (TSS)

Under the Shelterwood System, lianas and undergrowth vegetation are eliminated first. Unwanted species of middle size are then poisoned with sodium arsenite to allow more light to reach the lower levels. After four to six years, all marketable species are felled. TSS was developed to improve the regeneration of valuable species, particularly African mahogany, utile, sapele and iroko. The system originated in Malaysia and was introduced under the direction of Malaysian foresters in 1944 in Nigeria and in 1946 in Ghana – without success.

Enrichment Planting

After exploitation and clearing the undergrowth, seedlings of timber species (African mahogany, utile, gedu nohor, niangon, lovoa and bosse) are planted in parallel strips at regular intervals. In the Côte d'Ivoire, this system was widely used during the 1960s. Enrichment planting of natural forests had already been attempted during the 1930s in Nigeria and later in Ghana. In the 1960s, however, the system was abandoned by both countries as was TSS as well. Many of the young trees planted in strips were soon smothered by other vegetation and died.

not hesitate to establish plantations within forest reserves. In Nigeria, the forestry service must prove the socio-economic value of trees if it wishes to protect forest reserves against conversion to other forms of land-use [135]. The Subri Forest Reserve in Ghana was also transformed into a gmelina plantation, an act which was actually supported by the United Nations Development Programme (UNDP).

At the demand of local populations, which had never overcome the loss of their rights to land-use in forest reserves, the Taungya System was also introduced. It entails the combined plantation of timber and agricultural crops on cleared land. During the first two to three years, mixed crops are cultivated between the young trees. Afterwards, the farmer must forfeit his plot of land and leave it to the forester and his trees. The government retains property and timber rights. Practically speaking, the Taungya System was little different from the traditional practice of shifting agriculture in which farmed plots are left after some years to allow for natural regeneration [136]. With the introduction of the Taungya System, Ghana's entire forest plantation program attained only 60% success. It was difficult to supervise the system in the forest outside and the timber yield was accordingly poor [134]. Since the farmers have no share in the timber profits earned by the plantations, the Taungya System has also proved to be of limited success in Nigeria.

Aside from forest reserves, there will soon be no significant forested areas left in West Africa (Tab. 17). Official forest policy in Ghana and Nigeria requires that areas not within forest reserves be opened up to other functions, notably agricultural use. Unfortunately, a combined use for both agricultural and timber purposes has scarcely been considered up to now, although it would be the best way of achieving a sustainable and ecologically balanced land-use. Although agroforestry has become a popular term today, foresters – more than farmers – are skeptical of agroforestry methods. They would have to work with other timber species, use different logging practices and integrate the farmers into their work. Instead, forestry services limit themselves to the use of less efficient but better known methods of management in forest reserves.

Neither is there a guarantee that forest reserves will actually remain forest: It was necessary to declassify approximately one quarter of the original 33000 square kilometers of forest reserve in the Côte d'Ivoire. The government agency for reforestation SODEFOR, assigned with the demarcation of the reserves, discovered numerous cleared areas. Entire forest reserves had vanished and today only 24000 square kilometers of "permanent forest" remain [6]. In Nigeria, federalist attitudes prevent a uniform national forest policy. Forest reserves, which are practically the only forests left in the country today, fall under the jurisdiction of the separate states. They are not always able to prevent farmers and loggers from overstepping the law. In Nigeria, deforestation continues at a rate of 350–400 square kilometers a year and mainly affects forest reserves. Even well-stocked reserves are cleared to make way for plantations of oil palm, rubber, citrus fruits and cocoa although fallow land lies unused nearby [136].

Table 17
Forest reserves (forêts classées) in the rainforest zone of the most important West African timber export countries [6 and various sources].

	Forest reserves (km²)	Proportion of total closed forest area (1985)
Liberia	16647	22.4%
Côte d'Ivoire	24042	18.9%
Ghana	16788	20.5%
Nigeria	21221	15.8%

The bark of mahogany logs is often removed before loading. The dense network of timber routes and access roads as well as the numerous loading areas are no lesser cause of deforestation.

The Rise and Fall of Timber Export

Tropical timber became a viable alternative to European wood after World War II, when trade with East European countries ceased and even more so when timber became noticeably scarce in western and southern Europe. With the upward economic trend in industrialized European countries, the demand for tropical hardwood was concentrated almost exclusively on West Africa. At the time, economic theory projected endless growth and particularly American economists were in favor of mobilizing the rainforest's timber supply and investing the profits in other more productive branches of economy [8]. This attitude conflicted with the traditional principles of forest management, especially in the British colonies. But Europe's trade interests had long determined its relationship to West Africa and here again they proved to be stronger. Today, most of West Africa's natural forests are exploited. In the meantime, however, those countries' domestic demand for timber has increased. This, combined with the fact that profits from roundwood exports are skimmed off by trading companies and processing in Europe, has led many governments to set export restrictions. But the European market has proved

The false classification of roundwood for export is a common practice by which timber exploiters make illegal profits at the cost of the country.

to be less stable than before the recession of 1975. Although Europe still imports tropical roundwood almost exclusively from Africa, interest in sawnwood began to shift to Asia. The Africans were left in the cold.

Only four of West Africa's nine countries produce significant quantities of tropical timber. In order of increasing importance for today's export, they are: Nigeria, Liberia, Ghana and the Côte d'Ivoire.

Although Nigeria looks back on a century of forest management, today it is scarcely able to cover its domestic need for timber. The country's population is more than double that of the other eight West African states combined. And the oil boom of the mid-1970s further strengthened the domestic demand for wood and wood products. Nigeria began to import more timber products (logs, sawnwood, veneers, pulp and paper) than it exported. The foreign trade balance continued to worsen until, in 1980, Nigeria showed

a deficit of US$ 247 million spent for the import of wood products. The catastrophic situation had already led to a timber export ban in the 1970s. At the same time, logging activities were intensified in the remaining forests (Fig. p.198). But a program of reforestation was also begun and appears to have been successful as far as the trade balance is concerned. FAO statistics for 1985 show the deficit reduced by US$ 100 million [137]. It is questionable, however, whether many intact rainforest areas will survive in Nigeria.

Liberia's timber industry is of lesser importance. Although practically the whole country was once covered with rainforest, its exports consisted mostly of rubber and of iron ore extracted from the Nimba mountains. The first timber firm, the Maryland Logging Company, did not move into the southeastern corner of Liberia until 1965. Between 1977 and 1980, only 33 timber companies actively made use of their concessions although logging permits had been distributed over most of the country. In order to reduce logging activities and prevent the forests from quickly being creamed of their best specimens, the Liberian Forest Development Authority (FDA) limited annual exploitation within larger concessions to 4% of the total area [138]. Despite fears that Liberia may merely lie ten years behind the dramatic situation in the Côte d'Ivoire, it has seldom reached more than one tenth its neighbor's annual roundwood production.

Selling Out Ghana's Forests

Once Ghanaian tribal chiefs had been stripped of authority over the forests in their territory, the distribution of concessions was also centralized under the Ministry for Lands and Mineral Resources. Under Nkrumah, the government strived to promote small to middle-sized enterprises owned by Ghanaians and the distri-

The sapele trunk loaded on this small Ghanaian entrepreneur's truck equals the average yield of more than two hectares of rainforest. The loss of a corresponding area of forest is usually the consequence of his profit.

bution of timber concessions to foreign companies was thus reduced. One hundred concessions were granted to Ghanaians from 1961–71, only two to foreign companies. Until then, most timber rights had been in European hands but now the large foreign concessions had been divided and distributed among Ghanaian enterprises. The average size of a concession decreased from 686 square kilometers to a mere 41 square kilometers. By 1967, logging permits had been issued for 75% of the forest reserves and for all unreserved forests. The government's policy had drastic consequences. By 1970, the number of firms had risen from 121 to 361. More than half were subcontractors to the actual concession holder [139].

In spite of government loans to timber producers, many firms soon went bankrupt. A few lucky individuals, however, found their way from rags to riches. One example is Francis Osei Kyeremateng: "Franco" rented chain saws, one tractor

and a truck in order to fell timber under contract to large concessionaires and sell it to a sawmill in Kumasi. He used the profits to purchase his own tractor and two used timber trucks. In 1967, he was able to exploit more than 500 hectares in the Subim Forest Reserve. Additional profits allowed him to add a bulldozer, a second tractor and another truck to his enterprise. After developing trade relations with a Dutch partner who was prepared to invest in his machinery, Franco gained the government's trust and was issued an import license for spare parts. Up to then, he had only worked as a subcontractor to concession holders without timber rights of his own. Not until Acheampong came into office, whose government was considered corrupt throughout the country, did "Franco Timbers" receive its own concession – right in the middle of Bia National Park! In 1976, two thirds of the park had to be declassified to the status of game reserve in order to provide Franco Timbers and a number

The sawmill of Gliksten West Africa located near Sefwi Wiawso in western Ghana. It is powered by a steam engine – wood cuttings are burned for fuel.

of other small concessionaires with logging permits. The pressure on the remaining natural forest reserves had drastically increased.

But Ghana's government program to promote native enterprises did not succeed. Many of the small firms fought costs much too high for what they profited [139]. Few Ghanaians made their way to the top and could allow themselves the luxury of being chauffeured through the rainforest in a Mercedes like "Franco". Instead, more and more Lebanese moved into the timber industry, openly or by using small-scale Ghanaian enterprises as fronts.

Ghana's forest service, which had to deal with limited staff, funds and equipment in any case, stood before a particularly difficult task. The government's policy of promoting small-scale enterprises led to relatively small concession areas with more intensive logging activities over a wider total area thus making supervision difficult. At the demand of timber concessionaires,

all economic timber trees with diameters of at least 108 centimeters were allowed to be felled. The timber concessionaires had "convinced" the government that trees of greater diameter were "overmature" and would result in a deterioration of wood quality. For this reason, the felling cycle was also reduced from 25 to 15 years. The more liberal regulations were based on totally unfounded ideas – most rainforest trees reach much greater diameters without becoming "overmature" – and led to unnecessary destruction and damage. Ironically, large numbers of felled trunks rotted in the forest because there was no market for them [133].

The hoards of small-scale enterprises led to the downfall of many rainforest areas. Even in the forest reserves, loggers often did not follow the rules of selective exploitation and indiscriminately felled all economic species. Not much was left of the top canopy. Forest guards were bribed to label best quality trunks as second class in order to lower their export value. This reduced the country's hard currency income but the exporters had their European customers pay them the difference to a bank account abroad, an illegal practice still common in West Africa today.

Attempts at Rectifying Ghana's Timber Industry

Large timber companies, some of which had been active in Ghana since 1948, also suffered from the preference given to small-scale enterprises. Timber companies operating in the well-stocked moist semi-deciduous forests were especially hard hit by the government scheme: Gliksten West Africa, the Mim Timber Company and African Timber and Plywood (ATP). The Mim Timber Company had paid particular attention to exploiting its concession carefully and protecting it from slash-and-burn clearing. In the 1970s, Mim was nationalized like many other foreign firms but instead of being rewarded for its considerate forest utilization, it was quite poorly treated. Not only were the older timber companies now struggling with the recession, their import licenses for spare parts were also restricted. Mim even lost some of its timber rights to small-scale enterprises favored by Acheampong's government.

Today's government under Jerry Rawlings is trying to correct some of the mistakes made in the past; divided concessions have been returned to their original holders. In 1979, the decrease in timber production led to a log export ban for 14 species in order to promote the domestic timber processing industry, but also to encourage the exploitation of lesser known species such as pterygota, aningre, cardboard and the relatively common limba. Some of today's experts recommend returning to earlier methods of selective exploitation and a much longer felling cycle of 40 years [133]. But today, Ghana's forests consist almost only of forest reserves and even they

have been too severely depleted of commercial species to attract further large foreign investors. Afrormosia, a tree species of limited distribution in West Africa, is listed today as rare and endangered [140]. Half of Ghana's aformosia stock once occurred within Mim concessions.

New loans from the World Bank have tempted smaller British firms back to Ghana. They take over the management of timber companies like the Bibiani Logging Co. Ltd., improve their equipment and in return they can export timber to Great Britain [141]. Since 1983, Ghana's timber production and export have begun to rise from an all time low (Fig. p.198). The question is, for how long.

Overexploitation in the Côte d'Ivoire

The Côte d'Ivoire has been and continues to be by far Africa's most important exporter of tropical timber. Although this country with the liberal market economy on the Gulf of Guinea cannot yet compete with Southeast Asia's timber giants, in the "golden" timber years of the 1970s, the Côte d'Ivoire did export one fifth the volume of Indonesia's export. Indonesia, however, logs its forests much more intensely because of the frequent occurrence of the dipterocarp species. In 1985, one quarter the value of Africa's total timber products export was covered by the Côte d'Ivoire [137].

The Côte d'Ivoire has a younger history of forest management than the former British colonies. Timber export remained insignificant until the 1950s but after independence, it was boosted all the more to support the country's economic development. No where else in Africa was the rainforest opened up so quickly and in such an uncoordinated manner as in the Côte d'Ivoire. The production of industrial roundwood truly corresponded to the market – twenty years ago (1969), the Côte d'Ivoire exported nearly the entire quantity of timber felled. There were

Tropical timber production and export 1962–1987. The black lines indicate the total production of roundwood (logs). The broken lines indicate the export of round-wood, sawnwood and woodbased panels (expressed in roundwood equivalents). Source: FAO [137].

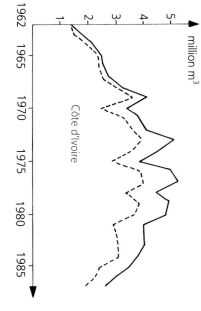

million m³

Côte d'Ivoire

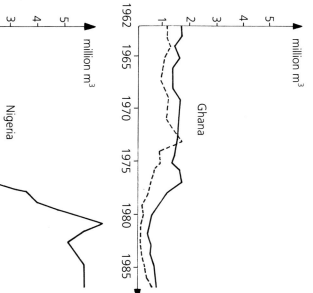

million m³

Ghana

million m³

Nigeria

d'Ivoire exported by far the majority of its timber in roundwood form. Only 30 years since the tropical timber boom really began, supplies are clearly showing signs of exhaustion (Fig. p.198). There is not much forest left to be opened up in the Côte d'Ivoire. The export of the more valuable species – African mahogany, utile, sapele, gedu nohor and makore is steadily decreasing in favor of the faster growing obeche, a species often taken in a second felling. In any case, the Côte d'Ivoire's liberal economy seems to be leading towards Nigeria's situation of 15 years ago. A country whose forest area consists of only small patches of secondary forest and whose timber exports threaten to collapse in the face of an increasing domestic demand for industrial timber.

The Legacy of Colonial Power

Arthur Creech Jones, a former Secretary of State for the British colonies, described Britain's well-meant colonial policy in West Africa early in the century: "To guide the colonial territories to responsible self-government within the Commonwealth in conditions that ensure to the people both a fair standard of living and freedom from oppression from any quarter". British colonies often refrained from establishing large plantations because it would have meant forcing the local inhabitants off their land. After their bad experiences with tribal chiefs in Ghana, the English sought to include the local chiefs in the planning process and to allow the village chiefs to rule their territory. In accordance with this colonial philosophy, school systems were encouraged and developed in British West Africa. Today, the level of education in Ghana and Nigeria remains superior to that in French-speaking West Africa.

France treated its colonies differently: French territory in West Africa was always considered part of the "Grande Nation". All property

2646 concessions distributed over 66000 square kilometers of forest, which is equal to the area of Sri Lanka [4]. Contrary to most other West African timber producers who sought to develop their own forest industry, the Côte

Wood cuttings being prepared for local use at the Mim Tiber Company.

automatically belonged to the state unless an African had registered his land with the French authorities. The French administered their territories by their own colonial officers and did not delegate any power whatsoever to the native people. Throughout French West Africa and especially in the Côte d'Ivoire, large tracts of land were given to foreign investors for the establishment of rubber, oil palm, pineapple and banana plantations. Business and trade were very much oriented toward the French market and remain so today. Former French colonies have continued to be greatly dominated by foreign powers even after attaining independence. After the Côte d'Ivoire became independent in 1960, it was not curious that the number of French citizens in the country had nearly doubled by 1972. In all English-speaking countries, on the other hand, the population of citizens belonging to the former colonial power had decreased. Today, English-speaking West Africa tends to disdain its French-speaking neighbors for their economic dependence on and factual control by the former colonial rulers. Professor of Geography Reuben K. Udo from the University of Ibadan (Nigeria), for example, comments that former French colonies need more export earnings to satisfy the demands for goods on the part of the expatriate population and their African elite: "Some of whom tend to be more French than the French". In spite of relatively impressive growth rates in the national economy, the Côte d'Ivoire's rural areas have remained as backward and underdeveloped as ever. Udo considers it an example of "growth without development" [2]. Whoever has seen crates of imported French table wine and Vichy mineral water standing beside clay huts on dusty roads will tend to share his opinion. Old ties to France have also developed into timber trade relations. Although the Côte d'Ivoire quickly rose to become Africa's most important timber exporter after World War II, not even the first step – sawing the trunks – was left

to the country of origin. The majority of Ivorian timber continues to be exported as roundwood logs to essentially the same destinations since the post-war timber crisis (Fig. p.200). Apart from France and Italy, Portugal is also an important consumer of Ivorian timber. Ivorian logs are used there to produce plywood and veneers which are then exported to other European countries, notably Britain! Since 1985, an increasing quantity of roundwood (saw logs and veneer logs) was exported from the Côte d'Ivoire to Japan. European countries, on the other hand, reduced their roundwood imports

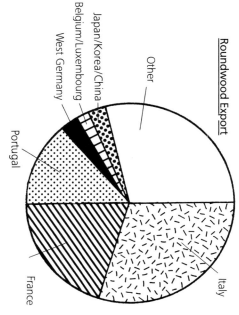

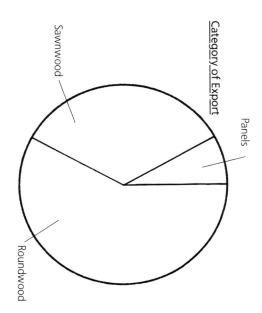

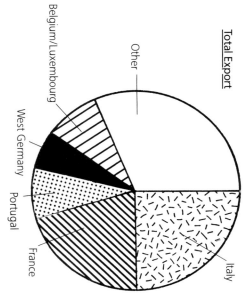

Category of Export

Panels

Sawnwood

Roundwood

Roundwood Export

Other

Japan/Korea/China

Belgium/Luxembourg

West Germany

Portugal

Italy

France

Total Export

Other

Belgium/Luxembourg

West Germany

Portugal

France

Italy

Export and destinations of tropical timber from the Côte d'Ivoire, 1985. Sawnwood and woodbased panels are expressed in roundwood equivalents. Source: FAO [137].

from the Côte d'Ivoire in favor of a higher proportion of sawnwood and panels. Originally, timber export in the Côte d'Ivoire had as elsewhere been the domain of European family firms, for example Victor Balet. Today, the enterprise consists of registered African companies based on African capital with timber rights, sawmills and other facilities in the Côte d'Ivoire and in the Central African Republic. Various large French firms are involved in the Ivorian timber business, BECOB, Lalanne, SCAF and SCOA among others. They in turn also control a number of Ivorian companies. And behind many a French industrial conglomerate stands a French bank like the BNP or Credit Agricole [142].

The Same Mistakes in Central Africa?

While English-speaking countries tended to nationalize the timber industry, transform traditional family-owned enterprises into semi-nationalized companies and limit or even ban export; the French merrily continued to dominate the forest industry in the Côte d'Ivoire – but not to the best interest of the forest. Ivorian roundwood production wavered with export demand which shows that logging did not follow the rules of sustainable forest management but, rather, those of the free market (Fig. p.198).

But the party is over in the Côte d'Ivoire as well. Timber production and export have dropped dramatically. Ivorian rainforests have been depleted and attention is slowly turning to Central Africa, where a great deal of rainforest still stands. Between 1981 and 1985, already two and a half times as much primary forest was opened up there as in West Africa (Tab. 15). Interestingly enough, industrial conglomerates tend to keep to their language areas more than anything else. While the British timber industry concentrates its efforts increasingly in Southeast

West Africa over the past 30 years will be repeated in Central Africa. Immigrant farmers have not as yet cleared as much land in the extensive forest areas of Central Africa as has been the case in West Africa, but the peculiarities of timber exploitation and the industry's methods have not changed. In 1978, Jack Westoby, a prominent forest economist at FAO, described the situation at the World Forestry Congress in Jakarta: "Forest industries have made little or no contribution to socio-economic development in the underdeveloped world – certainly not the significant contribution that was envisaged for them a couple of decades ago. Indeed, the probability is that such forest industries (...) served but to deflect attention from real needs, diverted resources from what should have been the true priorities, and served to promote socio-economic *underdevelopment*" [143]. Hansjürg Steinlin, Professor for global forestry at the University of Freiburg i.Br., still comes to the same conclusion a decade later. Steinlin states, "The capital made available by timber harvests was not reinvested in agriculture and forest management, instead it was withdrawn for the most part from the rural regions and trickled in sometimes dubious ways to the cities where it was spent on consumer goods or invested in unproductive ways. The costs to the national economy caused by forest loss are well above the profits" [8]. These are not the words of fundamentalist conservationists but the opinions of experienced tropical foresters who have the courage to speak out. Their statements are especially true in West Africa. In view of the catastrophic consequences of timber exploitation in closed tropical forests for the ecology, local populations and national economies; African governments, foreign investors, importer countries and consumers should today ask themselves if it might not be better to completely refrain from this form of rainforest utilization.

Asia, French firms do more business in Cameroon, Gabon and in the Central African Republic. But the large German Danzer-Group, active in tropical timber exploitation and processing worldwide, has also shifted its African business from the Côte d'Ivoire to Cameroon and particularly to Zaïre. West Africa has served its purpose – in return, the countries along the Gulf of Guinea will soon be able to puzzle over how to meet their own rising demands for industrial wood. So far, the timber plantation program has only produced significant yields in Nigeria.
The danger is great that the mistakes made in

Forest Conservation Past and Present

Forest inhabitants present the first sip of every calabash brimming with palm wine to their ancestors, allowing it to trickle into the ground as symbol of gratitude for the land and hunting grounds left to their use. Neither will a son unthinkingly sell his mother's hut. These people know the value of their old drums and of the forest which has supported them and their family for generations.

Tribal chiefs and village elders may have appeared to only be defending their rights to ancestral land as they resisted the centralization of the forest administration at the beginning of the century. But even then it was a question of conservation. Although the forest inhabitants were not familiar with the term, their cultural and existential dependency on the forest and its gifts was synonymous with conservation. Their life meant conservation. Of course, neither did the colonial governments, the newly established forest administrations nor even many timber companies intentionally plan to destroy the forest. On the contrary, the creation of forest reserves was meant to place as much forest as possible under permanent protection.

Paradoxically today, in spite of the fact that everyone strived towards forest conservation back then, or perhaps just because of that, exactly the opposite was achieved. The incompatibility be-

The throne of the "Chief" is taken from storage. Today, only a few people are interested in the village chief's wishes.

>> Representatives of the government authorities, the local population and conservation organizations participated in the planning seminar for the establishment of the first national park in Liberia, Sapo National Park. The seminar resulted in a widely supported management plan which is now in effect.

>> Jim Thorsell, Secretary of IUCN's Commission on National Parks and Protected Areas, tests the possibilities for non-destructive tourist activities on the Sinoe River along the border of Sapo National Park.

tween tradition and modern forest management has already been dealt with in the first chapter of this book. To put it concisely: The forest conservation planned on paper, without consideration of local interests or effective methods of supervision, and combined with increasing pressure from the timber industry and immigrant farmers, culminated in forest destruction. Or, in the words of forest economist Westoby: "Though every underdeveloped country now has a forest service, these forest services are nearly all woefully understaffed, and miserably underpaid. Because they exist, exploitation is facilitated; because they are weak, exploitation is not controlled. Because exploitation has been uncontrolled, and management non-existent, marginal farmers, shifting cultivators, and landless poor have followed in the wake of the loggers, completing the forest destruction" [143]. If the same mistakes are not to be repeated, forest conservation measures must be conceived with respect for local interests.

From Hunting Bans to Agroforestry

Forest management introduced conservation measures more to preserve the timber supply than to conserve biological diversity as a whole. And 20 years ago, on the other hand, conservationists gave little thought to the fate of West African rainforests as an ecosystem. They were mainly concerned with aspects of species conservation and tried to preserve animals from being hunted, irrespective of rising threats to the forest as a whole. Even in the 1960s, it was apparently hard to imagine that mere patches of forest were soon to remain of West Africa's rainforests.

Conservationists made their recommendations accordingly. They modelled hunting after European traditions, with open and closed seasons, bans on certain species or on females and young offspring, and sometimes went as far as

An example of traditional agroforestry: oil palms, plantains and coffee shrubs in a thinned out patch of forest in western Cameroon.

to completely ban hunting [144, 145]. The use of traps and snares, the trade and even transport of bushmeat were also examined and actually prohibited in some West African countries. Of course, hunting regulations can be useful if local needs are taken into account and if the regulations can be enforced. But how is a village hunter to determine between a male or a female duiker antelope in the rainforest, and judge its age, if he often cannot even distinguish between an antelope and a monkey in the light of his carbide lamp? And when people cover their protein needs almost exclusively with bushmeat,

what are they supposed to eat if hunting is suddenly not allowed for months on end? It is not surprising that West African hunting regulations have remained nothing but worthless pieces of paper and are scarcely respected anywhere. First, the Europeans deprived the forest inhabitants of their authority over the forest in the belief of so being able to better protect nature. Then, they tried to preserve wildlife resources with European-style hunting regulations. It would hardly have been possible to attempt more unsuitable methods of forest conservation and they were consequently of little success.

International Efforts

Today, there is a global demand for land-use planning in tropical rainforest areas. Sustainable, non-destructive methods of utilization should be the basis of such planning. But how can that demand be met when we still do not know how rainforests can be managed sustainably and when foresters knowingly cling to old methods despite their unsuitability?

The Tropical Forestry Action Plan conceived for 1987–91 by the World Resources Institute in cooperation with the Food and Agriculture Organization (FAO), the United Nations Development Programme (UNDP) and the World Bank with an initial budget of US$ 5.3 billion was developed with high hopes [146]. Saving the world's forests was assumed to be a costly undertaking. Through increased investments in forestry, it was believed that the destruction could be halted. The Action Plan has since been judged more objectively. Forestry review reports of missions to various tropical forest countries do not look particularly promising. Although firewood and industrial timber plantations were included, old methods of timber exploitation were not questioned and a mere 8% of the budget was reserved for conservation projects. Significantly, the large teams of forestry experts who judged the situation in Ghana, the Côte d'Ivoire and in Cameroon scarcely included critical foresters and only a few forest conservationists.

The International Tropical Timber Organization (ITTO) also has yet to prove that its activities actually result in forest conservation. A part of its efforts aims at financing forest conservation projects under article 1(H) of the International Tropical Timber Agreement (ITTA) established in 1984. In the long term, however, the treaty could prove to be an effective instrument for international forest conservation. It is the only commercial treaty to even mention tropical forest con-

servation, let alone declare it an aim. A further advantage of the ITTO is the equal participation of producer and consumer countries which enables critical voices in the tropical timber debate to also directly influence timber exploitation.

Budowski's Principles of Forest Conservation

International efforts to conserve tropical forests should not be overestimated. Especially not when new investments in the forest sector are supposed to correct what previous investments have destroyed. In view of today's continuing forest loss, not only are new methods needed but an entire new philosophy is necessary based on the urgent needs of conservation as well as the social and economic needs of today's forest inhabitants. In many rainforests throughout the world – and very much so in West Africa – preserving remaining undisturbed forest has to be the leitmotif of every planning concept. Forestry, too, must adhere to this principle. In other words: Forest utilization planning needs to concentrate on areas already opened up – secondary forests and fallows. In 1984, one of the most progressive tropical foresters, Gerardo Budowski, based his principles of forest utilization on just this need [147]:

a) *Agroforestry*: Integration of timber, firewood and fruit trees in agricultural plots based on traditional and modern methods.

b) *Plantation of trees*: Reforestation with native and exotic species for the production of timber, pulp and fuelwood but never at the cost of natural forests. There is enough fallow land to be used for such purposes.

c) *Secondary forest management*: Production of valuable timber species in already existing secondary forests, using methods of sustained production such as liberation cuttings.

Taï National Park comes under heavy pressure from new settlers. In swampy areas, even rice is cultivated (photo above). The park's few employees can scarcely deal with the pressure for land around West Africa's largest national park (photo below).

Like many forest waterways in West Africa, this small river inside Taï National Park carries gold. Gold washers move in illegally to prospect and largely live off wild game.

d) *Social forestry*: Production of timber, fuelwood, fibers, medicinal plants and charcoal by village communities, on a sustained basis also as part of agroforestry systems.

e) *Buffer zones*: Management of forest zones surrounding national parks and other protected areas. Such areas provide an opportunity for foresters and conservationists to prove their cooperation.

Protected forest areas are especially suitable for focusing the development of integral forest utilization. Inversely, a careful, sustainable land-use in zones surrounding protected areas is of vital importance to the areas themselves if they are not to be sacrificed to an increased demand for land.

Rainforest National Parks in West Africa

Until 1968, there was not a single strictly protected rainforest area in all of West Africa, with the exception of the 180 square kilometer large nature reserve in the Nimba mountains and the tiny Banco National Park northwest of Abidjan (Fig. pp.212–13). Later, Mont Peko National Park and Marahoué National Park were established in the Côte d'Ivoire. The latter comprises over 1000 square kilometers of transitional zone to Guinea savannah. In 1972, Taï National Park was established in the southwestern Côte d'Ivoire at the recommendation of the International Union for Conservation of Nature (IUCN) and WWF [151]. The Taï Park – with an area of 3300 square kilometers, a buffer zone and the neighboring N'Zo Wildlife Reserve – remains the most important protected rainforest area in West Africa today. Although it has not been

Undisturbed rainforest covers the highest point of elevation in Korup National Park. Due to its mountainous topography, timber exploiters have scarcely ventured into this area of Cameroon's forest.

>>
At the entrance to a village settlement in the Korup area. The village is two days' walking distance from the nearest road. Footpaths through closed rainforests provide the only connection with the outside world.

overly careful of its forests, one must admit that Africa's "land of economic wonder" did set aside integrally protected areas earlier and in a more consistent manner than any other country along the Gulf of Guinea. Until very recently, Sierra Leone, Guinea and Nigeria had not managed to establish even one national park in the rainforest zone. The mangrove forests in Guinea Bissau, so very important for the fishing industry, were virtually unprotected. In 1989, however, a coastal wetlands management project was begun there. Within the scope of this project, a series of reserves are to be established which will allow multiple use of Guinea Bissau's coastal wetland

resources. Also in 1989, the Nigerian Council of Ministers upon the recommendation of WWF approved two important rainforest areas in southeast Nigeria being declared a national park. The Cross River National Park will incorporate the Oban Division, a large area between the Cross River and the Cameroon border, and the Boshi-Okwangwo Division, a geographically separate forest area further north (Fig. pp.212/13). Including these new additions to the West African system of protected areas, only about 10500 square kilometers have national park or strict nature reserve status in the rainforest zone. An additional 3660 square

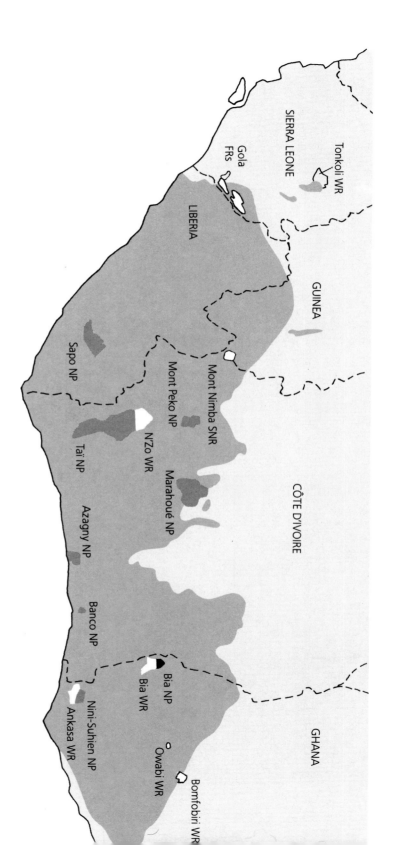

kilometers are classified as wildlife reserves. These figures comprise only 2.7% of the total rainforest area. An additional area of about 10000 square kilometers has been suggested for protection [149]. The future reserve areas planned in Sierra Leone and Liberia, however, have been on a waiting list for some time and it is doubtful whether they will ever receive protected status. But whatever the case, the total protected rainforest area by far does not correspond to the importance that should be attributed to preserving biological diversity and the ecological balance in West Africa. Of course, the forest reserve system is also vital (Tab. 17), but as was mentioned earlier, many of the forest reserves suffer from illegal slash-and-burn clear-

ing. Others have been exhausted by hunting or have even been transformed into timber plantations. High priority must therefore be given to creating additional strictly protected forest areas in West Africa.

Threats to National Parks in the Côte d'Ivoire and Ghana

Uncoordinated and uncontrolled land-use almost had fatal consequences for West Africa's largest protected rainforest area, the 3300 square kilometer large Taï National Park in the southwestern Côte d'Ivoire. The immigration of farmers from the Sahel region as of 1980 led to ever-increasing slash-and-burn activities in the

Protected areas in West Africas rainforest zone. This sketch includes national parks (NP, dark areas), strict nature reserves (SNR) and wildlife reserves (WR). Source: IUCN and others [148, 149, 150]. Forest reserves and some other reserves heavily damaged by timber exploitation or human settlement are not shown. Only the Gola Forest Reserves (Sierra Leone) are additionally indicated since they are of special significance and may soon receive a higher protection status.

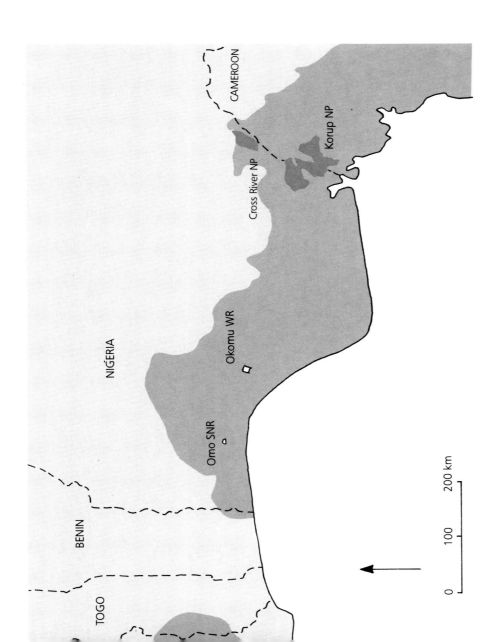

immediate surrounding areas of the park. At the same time, wildlife came under increased hunting pressure. And at times, several hundred gold washers were present inside the park's boundaries. Timber companies also overstepped their rights in the northern areas. Aerial photographs of Taï National Park's border areas made in 1988 showed that farmers and timber exploiters did not encroach upon the buffer zone where it was clearly marked by a peripheral road. Clear demarcation is a simple but often crucial method of forest protection. Many forest reserves in West Africa have been sacrificed for want of clearly marked boundaries or because these were overgrown. An agreement between the Ivorian government and WWF on the manage-

ment of the Taï region was signed in 1988 and obliges both parties to financially support the maintenance of Taï National Park. But it will not be easy to compensate for the earlier lack of land-use planning, nor to stabilize the situation which has resulted from destructive land-use around one of the most valuable rainforest areas in West Africa.

One of the former Chief Game and Wildlife Officers in Ghana, Emmanuel O. A. Asibey, was the first to press for protected areas representing all of the country's vegetation zones. Ghana possesses an extensive system of forest reserves but they mainly serve the timber industry [152]. Bia National Park was established in 1974 and only two years later, it was reduced

Sapo National Park (Liberia) with centers of sustainable agricultural development (squares).

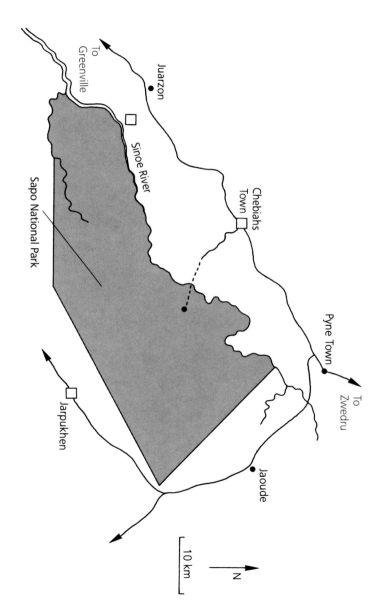

to a minimal 77 square kilometers under the pressure of several small timber enterprises. The remaining 228 square kilometers were declassified to game reserves and partially opened up to timber exploitation. In the same year, 1976, this misfortune was compensated by establishing Nini Suhien National Park and Ankasa Game Reserve in the moist southwest corner of the country (Fig. pp.212/13). Together, they comprise 513 square kilometers. These areas suffer less pressure from the timber industry ever since the failure of the firm George Grant due to a lack of marketable timber species. In 1989, a further rainforest area in southern Ghana, the Kakum Forest Reserve, was suggested to be established as a national park.

Planning Sapo National Park

Liberia established its first strictly protected rainforest area in 1983. Sapo National Park, which

covers an area of 1307 square kilometers, is still the only area of this status in the country (Fig. p.214). In 1978, Jacques Verschuren was assigned by WWF and IUCN to make recommendations for wildlife conservation and national parks based on a survey of the possibilities throughout the country [153]. But the true responsibility for the establishment of Sapo National Park lies with Alexander Peal, Director of Liberian Wildlife and National Parks. Although timber concessions had already been distributed in the area and specific trees had already been marked for felling, he continued to strive patiently towards his aim. In order to secure the widest support possible for the protection and management of Sapo National Park, a seminar was held in 1985 in which representatives of the concerned ministries, the local population, WWF and IUCN participated. It was the beginning of the first integral management plan in West Africa. Besides the management of the park, the

Korup National Park (Cameroon) with surrounding rural development zone. This area includes many forest villages which can be reached only by footpaths.

> No pesticides are used for cocoa cultivation in the isolated areas surrounding Korup National Park. Small-scale farmers have to head-carry their products to the nearest road for transport.

215

plan prescribed land-use and development in the surrounding areas. At first, local farmers suffered from fluctuating and uncertain harvests and a lack of workers. The project now includes the development of sustainable agriculture.

Preserving the Balance in Korup

Today, the needs of the local population are taken into much greater consideration in planning new protected areas. In 1986, the establishment of Korup National Park placed the park's 1259 square kilometers within the framework of a comprehensive land-use concept (Fig. p.215). The Korup area supposedly consists of the oldest rainforests in Africa. Situated at the edge of the center of endemism "West Central" (Fig. p.39) in the wettest area of the continent, Korup harbors an accordingly high diversity of plant and animal species. Across the border in Nigeria, the recently established Cross River National Park belongs to the same forest area. Thus, the entire southern border area between Nigeria and Cameroon has doubtlessly become a protected rainforest complex of global importance (Fig. pp.212/13).

Hilly terrain and deep gorges have hindered the opening up of the Korup area. The population density has remained low despite patches of fertile basalt soil outside Korup National Park. Only the Pamol (Unilever) oil palm plantation had attracted people from other regions to settle in the humid southwest and eventually led to an increased demand for land. Forest villages in the Korup area can be reached only on foot along narrow paths through closed rainforest and are thus practically isolated from what we tend to call civilization. They are often surrounded by small-scale farms where wonderfully mixed plots of plantains, cocoyams, cassava and vegetables are cultivated together with cocoa and coffee. There are often trees left standing within the plots. Tree species of the *Mimosaceae* family, *Albizia* spp. for example, seem to be favorites with the farmers. Their feathery leaves cast partial shade on the plots and probably fix nitrogen from the air which they then pass on to the soil. Some of the traditional farms in the Korup area are pefect examples of agroforestry and could serve as models for other regions.

WWF has negotiated a complex and ambitious project with the government of Cameroon to ensure the balance between nature conservation and land-use. The project is cofinanced by the British Overseas Development Administration (ODA) and German development aid. Besides national park management, the program includes different aspects of natural resource conservation in the surrounding forests ranging from agroforestry and livestock holding to hunting. Research, education and the development of a well-adapted form of tourism are also included in the project.

In order to plan future land-use in areas surrounding Korup National Park, a soil fertility survey was conducted [154] as well as a socio-economic survey of the forest inhabitants and a study of their hunting and trapping methods [120, 155]. The knowledge so gained aids in retaining the ecological balance of the Korup area and improving the situation in favor of the local population. And just this is of importance for roads near Korup National Park have been planned for years now which could endanger what has thus far been achieved. But it is not easy to make decisions, nor to put them into effect. The local village councils must agree with whatever is decided and sometimes they would simply prefer a road to finally put them in touch with the rest of the world.

Lessons for the Future

Land-use plans are only useful if they are feasible and supported by the local population. In

There are still forest inhabitants in West Africa who know how the rainforest can be used without destroying it. Their villages may seem primitive but appearances deceive. The people here could be the key to forest utilization in the future.

this context, the experience gained with the establishment of forest reserves earlier this century should be a lesson for the future. In the wake of today's forest loss, many West African governments are considering careful land-use in regions surrounding the remaining forest areas. And at least farsighted politicians and administrators are willing to work toward integral forest conservation. These officials have to be supported with moral and, even more importantly, material aid, in order for them to put their costly plans into action. Otherwise, industrialized countries will miss their last chance to save West African forests.

Development aid agencies are becoming increasingly interested in forest conservation projects which take the needs of forest inhabitants into account. But Africa is only beginning to see the truly sustainable use of rainforest today. Successful projects are urgently needed to serve as models for governments and investors. In addition, forest policy must refrain from the uncoordinated, large-scale opening up of still more rainforest. The countries of Central Africa with their vast expanses of forest would by all means do well to learn from the past decades of West African history.

References

1 Dickson, K. B. 1969. A historical geography of Ghana. Cambridge University Press, 379 pp.

2 Udo, R. K. 1978. A comprehensive geography of West Africa. Heinemann, Ibadan.

3 Oliver, R. and J. D. Fage. 1975. A short history of Africa, 5th ed. Penguin Books.

4 Arnaud, J. C. et G. Sournia. 1980. Les forêts de Côte d'Ivoire. Ann. Univ. Abidjan, série G, IX: 6–93.

5 Unwin, A. H. 1920. West African forests and forestry. London.

6 FAO/UNEP. 1981. Tropical forest resources assessment project: Forest resources of tropical Africa. Part I: Regional synthesis. Part II: Country briefs. FAO, Rome.

7 Martin, C. 1985. West- und zentralafrikanische Regenwälder: Kaum genutzt und doch zerstört. In: Kahlschlag im Paradies, P. E. Stüben (ed.). Focus Verlag, Giessen, 103–121.

8 Steinlin, H. 1988. Tropische Waldnutzung – Raubbau oder nachhaltige Forstwirtschaft? In: Tropische Regenwälder, E. Bugmann (ed.). Ostschweiz. Geogr. Ges. St. Gallen: 29–59.

9 CCTA/CSA. 1956. Phytogéographie / Phytogeography. Publ.no. 53.

10 White, F. 1983. The vegetation of Africa: A descriptive memoir to accompany the Unesco/AETFAT/UNSO vegetation map of Africa. Unesco, Paris.

11 Knapp, R. 1973. Die Vegetation von Afrika. G. Fischer Verlag, Stuttgart.

12 Sommer, A. 1976. Attempt at an assessment of the world's tropical moist forests. Unasylva 28: 5–24.

13 Hamilton, A. C. 1982. Environmental history of East Africa. A study of the quaternary. Academic Press, London.

14 Livingstone, D. A. 1982. Quaternary geography of Africa and the refuge theory. In: Biological diversification in the tropics by G. T. Prance (ed.). Columbia Univ. Press.

15 Hamilton, A. 1976. The significance of patterns of distribution shown by forest plants and animals in tropical Africa for the reconstruction of upper Peistocene palaeoenvironments: A review. In E. M. van Zinderen Bakker (ed.): Palaeoecology of Africa, the surrounding Islands and Antarctica. 9: 63–97.

16 Booth, A. H. 1958. The Niger, the Volta and the Dahomey Gap as geographic barriers. Evolution 12: 48–62.

17 Grubb, P. 1982. Refuges and dispersal in the speciation of African forest mammals. In: Biological diversification in the tropics by G. T. Prance (ed.). Columbia Univ. Press.

18 Hall, B. P. and R. E. Moreau. 1970. An atlas of speciation in African passerine birds. Trustees of the British Museum (Natural History). London.

19 Guillaumet, J. L. 1967. Recherche sur la végétation et la flore de la region du Bas-Cavally (Côte d'Ivoire). Mem. OSTROM 20: 1–247.

20 Haffer, J. 1982. General aspects of the refuge theory. In: Biological diversification in the tropics by G. T. Prance (ed.). Columbia Univ. Press.

21 Selander, R. K. 1971. Systematics and speciation in birds. In: Avian biology by D. S. Farner and J. R. King (eds.) 1: 57–147. Academic Press, New York.

22 Schiøtz, A. 1967. The treefrogs (Rhacophoridae) of West Africa. Spolia zoologica musei hauniensis 25: 1–346.

23 Flohn, H. 1968. Vom Regenmacher zum Wettersatelliten. Klima und Wetter. Kindler, München.

24 Ojo, O. 1977. The climates of West Africa. Heinemann, London.

25 Jenik, J. and J. B. Hall. 1966. The ecological effects of the harmattan wind in the Djebobo Massif (Togo Mountains, Ghana). J. Ecol. 54: 767–79.

26 Longman, K. A. and J. Jenik. 1987. Tropical forest and its environment, 2nd ed. Longman, London.

27 Hall, J. B. and M. D. Swaine. 1976. Classification and ecology of closed-canopy forest in Ghana. J. Ecol. 64: 913–951.

28 Hall, J. B. and M. D. Swaine. 1981. Distribution and ecology of vascular plants in a tropical rain forest. Forest vegetation in Ghana. W. Junk Publishers, The Hague.

29 Dahms, K. G. 1979. Afrikanische Exporthölzer, 2. ed. DRW-Verlag, Stuttgart.

30 Taylor, C. J. 1952. The vegetation zones of the Gold Coast. For. Dep. Bull. 4: 1–12.

31 Guillaumet, J. L. et E. Adjanohoun. 1971. La végétation de la Côte d'Ivoire. Dans: Le milieu naturel de la Côte d'Ivoire. Mémoires ORSTOM 50: 157–264, avec carte 1:500000.

32 Hall, J. B. (Nigeria) 1977. Forest-types in Nigeria: An analysis of pre-exploitation forest enumeration data. J. Ecol. 65: 187–199.

33 Hutchinson, J., J. M. Dalziel and R. W. J. Keay. 1954–72. Flora of West tropical Africa (2nd ed.). Crown Agents for Oversea Governments and Administrations, London.

34 Richards, P. W. 1973. Africa, the "odd man out". In Meggers (ed.), Tropical forest ecosystems in Africa and South America: A comparative review. Smithsonian Institution, Washington, pp. 21–26.

35 Jacobs, M. 1988. The tropical rain forest. Ed. by R. Kruk et al. Springer Verlag Berlin, Heidelberg. 295 pp.

36 Poore, M. E. D. 1968. Studies in Malaysian rain forest. The forest on triassic sediments in Jengka Forest Reserve. J. Ecol. 56: 143–196.

37 Hall, J. B. 1978. Check-list of the vascular plants of Bia National Park and Bia Game Production Reserve. In: C. Martin, Management Plan for the Bia Wildlife Conservation Areas, Part I. Wildlife and National Parks Division, Ghana Forestry Commission. Final Report IUCN/WWF Project 1251.

38 Thomas, D. W. 1986. The botanical uniqueness of Korup and its implications for ecological research. Proc. Workshop on Korup National Park (J. S. Gartlan and H. Macleod eds.). WWF/IUCN Project 3206.

39 McGregor-Reid, G. 1988. The fishes and the rainforest. Preliminary investigation, report and recommendations on the fish and fisheries of the Korup rainforest, West Cameroon. WWF publication Project 3206.

40 Anon. 1988. Monkey makes its debut. New Scientist, June 23, 1988: p. 31.

41 Collins, N. M. 1983. Termite and their role in litter removal in Malaysian rain forests. In: Tropical rain forest: Ecology and management, by S. L. Sutton, T. C. Whitmore and A. C. Chadwick (eds.). Blackwell Scientific Publ., Oxford.

42 Anderson, J. M. and M. J. Swift. 1983. Decomposition in tropical forests. In: Tropical rain forest: Ecology and management, S. L. Sutton, T. C. Whitmore and A. C. Chadwick (eds.). Blackwell Scientific Publ., Oxford.

43 Janzen, D. H. 1976. Why tropical trees have rotten cores. Biotropica 8: 110.

44 Irvine, F. R. 1961. Woody plants of Ghana. Oxford University Press, London.

45 Johansson, D. 1974. Ecology of vascular epiphytes in West African rain forest. Acta phytogeogr. Suec. 59: 1–29.

46 Rahm, U. 1972. Zur Verbreitung und Oekologie der Säugetiere des afrikanischen Regenwaldes. Acta Trop. 29/4: 452–473.

47 Schiøtz, A. and H. Volsøe. 1959. The gliding flight of Holaspis guentheri Gray, a West African lacertid. Copeia, 3: 259–60.

48 Wolf, E. C. 1987. On the brink of extinction: Conserving the diversity of life. Worldwatch Paper 78, Worldwatch Institute, Washington.

49 Collins, M. and S. Wells. 1987. Conserving invertebrates. IUCN special report compiled at the Conservation Monitoring Centre, Cambridge. IUCN Bulletin 18/7–9.

50 Collins, M. and M. G. Morris. 1985. Threatened swallowtail butterflies of the world. IUCN Red Data Book. IUCN Gland and Cambridge, 401 pp.

51 Topoff, H. 1984. Social organisation of raiding and emigration in army ants. Behaviour 14: 81–126.

52 Cansdale, G. S. 1961. West African snakes. West African Nature Handbooks, Longman, London.

53 Erwin, T. L. 1982. Tropical forests: their richness in Coleoptera and other arthropod species. Coleopt. Bull. 36: 74–75.

54 Erwin, T. L. 1983. Beetles and other insects of tropical forest canopies at Manaus, Brazil, sampled by insecticidal fogging. In: Tropical rain forest: Ecology and management, by S. L.Sutton, T. C. Whitmore and A. C. Chadwick (eds.). Blackwell Scientific Publications, Oxford.

55 Greenwood, S. R. 1987. The role of insects in tropical forest food webs. Ambio 16/5.

56 Boorman, J. 1970. West African butterflies and moths. West African Nature Handbooks. Longman, London.

57 Holden, M. and W. Reed. 1972. West African freshwater fish. West African Nature Handbooks. Longman, London.

58 Schiøtz, A. 1963. The amphibians of Nigeria. Vidensk. Medd. fra Dansk naturh. Foren. 125:1–92.

59 Waitkuwait, W. E. 1981. Untersuchungen zur Brutbiologie des Panzerkrokodils (Crocodilus cataphractus) im Taï Nationalpark in der Republik Elfenbeinküste. Diplomarbeit, Ruprecht-Karls Universität, Heidelberg. Mimeo.

60 Waitkuwait, W. E. 1988. Untersuchungen zur Erhaltung und Bewirtschaftung von Krokodilen in der Republik Côte d'Ivoire. Diss. Ruprecht-Karls Universität, Heidelberg. Mimeo.

61 Amadon, D. 1973. Birds of the Congo and Amazon forests: A comparison. In: Meggers (ed.), Tropical forest ecosystems in Africa and South America: A comparative review. Smithsonian Institution, Washington, pp. 267–277.

62 Chapin, J. P. 1932–54. Birds of the Belgian Congo. Bull. Am. Mus. Nat. Hist. Vols. 65 (1932), 75 (1939), 75A (1953), 75B (1954).

63 Mackworth-Praed, C. W. and C. H. B. Grant. 1970–73. Birds of West Central and Western Africa. Vol. I (1970), Vol. II (1973). Longman, London.

64 Moreau, R. E. 1966. The bird faunas of Africa and its islands. Academic Press. New York and London.

65 Davies, A. G. 1987. The Gola forest reserves, Sierra Leone. Wildlife conservation and forest management. IUCN Tropical Forest Programme. IUCN Gland, Switzerland and Cambridge, UK.

66 Martin, C. 1982. Management plan for the Bia Wildlife Conservation Areas, part I. Wildlife and National Parks Division, Ghana Forestry Commission. Final report IUCN/WWF project 1251.

67 Gartlan, J. S. 1984. The Korup Regional Management Plan (draft). Publication 25–106, Wisconsin Regional Primate Research Center.

68 Lang, E. M. 1975. Das Zwergflusspferd (Choeropsis liberiensis). Neue Brehm Bücherei, Wittenberg Lutherstadt.

69 Büttikofer, J. 1890. Reisebilder aus Liberia. Leyden.

70 Roth, H. H., M. Mühlenberg, P. Roeben und W. Barthlott. 1979. Gegenwärtiger Status der Comoé- und Taï Nationalparks sowie des Azagny-Reservates und Vorschläge zu deren Erhaltung und Entwicklung zur Förderung des Tourismus. Band III: Taï-Nationalpark. PN 73.2085.6, FGU Kronberg.

71 Roth, H. H. und G. Merz. 1986. Vorkommen und relative Häufigkeit von Säugetieren im Taï-Regenwald der Elfenbeinküste. Säugetierkundl. Mitteilungen 33: 171–193.

72 Bourlière, F. 1973. The comparative ecology of rain forest mammals in Africa and tropical America: Some introductory remarks. In: Meggers (ed.), Tropical forest ecosystems in Africa and South America: A comparative review. Smithsonian Institution, Washington, pp. 279–292.

73 Medway, L. 1978. The wild mammals of Malaya (Peninsular Malaysia) and Singapore. 2nd ed. Oxford University Press.

74 Gartlan, J. S., D. B. McKey, P. G. Waterman, C. N. Mbi and T. T. Struhsaker. 1980. A comparative study of the phytochemistry of two African rain forests. Biochem. Systematics and Ecol. 8: 401–422.

75 Waterman, P. G. 1983. Distribution of secondary metabolites in rain forest plants: toward an understanding of cause and effect. In: Tropical rain forest: Ecology an management, S. L. Sutton, T. C. Whitmore and A. C. Chadwick (eds). Blackwell Scientific Publications, Oxford.

76 McKey, D. B., P. G. Waterman, C. N. Mbi, J. S. Gartlan and T. T. Struhsaker. 1978. Phenolic content of vegetation in two African rain forests: ecological implications. Science 202: 61–64.

77 Rosevear, D. R. 1969. The rodents of West Africa. London. 604 pp.

78 Struhsaker, T. T. and J. F. Oates. 1975. Comparison of the behaviour and ecology of red colobus and black and white colobus monkeys in Uganda: a summary. In: Socioecology and psychology of primates, R. H. Tuttle (ed.). The Hague.

79 Struhsaker, T. T. 1979. Socioecology of five sympatric monkey species in the Kibale forest of Uganda. Advances in the study of behaviour 9: 159–228.

80 Olson, D. K. 1980. Male interactions and troop split among black- and white colobus monkeys (Colobus polykomos vellerosus). Paper presented at the VIIth Congr. of the Int. Primat. Soc., Florence, Italy.

81 Martin, C. and E. O. A. Asibey. 1979. Effect of timber exploitation on primate population and distribution in the Bia rain forest area of Ghana. Paper presented at the VIIth Congr. of the Int. Primatol. Soc., Bangalore, India.

82 Hoppe-Dominik, B. 1984. Etude du spectre des proies de la panthère, Panthera pardus, dans le Parc National de Taï en Côte d'Ivoire. Mammalia 48/4: 477–487.

83 Rucks, M. 1976. Notes on the problems of primate conservation in Bia National Park, Ghana. Dept. Game and Wildlife, unpublished report.

84 Olson, D. K. and S. Curtin. 1984. The role of economic timber species in the ecology of black-and-white colobus and diana monkeys in Bia National Park, Ghana. Abstract. Int. J. Primatology 5/4: 371.

85 Struhsaker, T. T. and P. Hunkeler. 1971. Evidence of tool-using by chimpanzees in the Ivory Coast. Folia primat. 15: 212–219.

86 Boesch, C. and H. Boesch. 1983. Optimisation of nut-cracking with natural hammers by wild chimpanzees. Behaviour 83 (3/4): 265–286.

87 Davies, A. G. 1987. Conservation of primates in the Gola forest reserves, Sierra Leone. Primate Conservation, The newsletter and Journal of the IUCN/SSC Primate Specialist Group 8: 151–153.

88 Roth, H. H. et B. Hoppe-Dominik. 1987. Répartition et statut des grandes espèces de mammifères en Côte d'Ivoire. IV Buffles. Mammalia 51/I: 89–109

89 Ehrlich, P. R. and P. H. Raven. 1964. Butterflies and plants: a study in coevolution. Evolution 18: 586–608.

90 Hocking, B. 1975. Ant-plant mutualism: evolution and energy. In: L. E. Gilbert and P. H. Raven (eds.), Coevolution of animals and plants. Univerity of Texas Press, Austin.

91 Bequaert, J. 1922. Ants in their diverse relations to the plant world. In: Wheeler, W. M., Ants of the Belgian Congo. Bull. Am. Mus. Nat. Hist. 45: 333–583.

92 Agland, P. 1982. Korup, an African rainforest film.

93 Jolivet, P. 1986. Les fourmis et les plantes. Un example de coevolution. Boubée, Paris. 254 pp.

94 Dalziel, J. M. 1955. The useful plants of West tropical Africa. Crown Agents for Oversea Governments and Administrations. Second reprint, London.

95 Janzen, D. H. 1972. Protection of *Barteria* (*Passifloraceae*) by *Pachysima* ants (*Pseudomyrmecinae*) in a Nigerian rain forest. Ecology. 53/5: 885–892.

96 Temple, S. A. 1977. Plant-animal mutualism: Coevolution with dodo leads to near extinction of plant. Science 197: 885–886.

97 Howe, H. F. and J. Smallwood. 1982. Ecology of seed dispersal. Ann. Rev. Ecol. Syst. 13: 201–28.

98 Gautier-Hion, A. et al. 1985. Fruit characters as a basis of fruit choice and seed dispersal in a tropical forest vertebrate community. Oecologia 65: 324–337.

99 McKey, D. 1975. The ecology of coevolved seed dispersal systems. In: L. E. Gilbert and P. H. Raven (eds.), Coevolution of animals and plants. University of Texas Press, Austin.

100 Snow, D. W. 1965. A possible selective factor in the evolution of fruiting seasons in tropical forest. Oikos 15: 274–281.

101 Haltenorth, Th. und H. Diller. 1977. Säugetiere Afrikas und Madagaskars. BLV, München.

102 Merz, G. 1982. Untersuchungen über Lebensraum und Verhalten des afrikanischen Waldelefanten im Taï-Nationalpark der Republik Elfenbeinküste unter dem Einfluss der regionalen Entwicklung. Diss., Heidelberg.

103 Sikes, S. K. 1971. The natural history of the African elephant. Weidenfeld und Nicholson, London.

104 Elder, W. H. and D. H. Rodgers. 1974. Immobilization and marking of African elephants and the prediction of body weight from foot circumference. Mammalia 38/I: 33–53.

105 Kurt, F. 1974. Remarks on the social structure and ecology of the Ceylon elephant in the Yala National Park. Int. Symp. on the behaviour of ungulates and its relation to management. Calgary. IUCN Publ. new series 24: 618–634.

106 Short, J. 1981. Diet and feeding behaviour of the forest elephant. Mammalia 45/2: 178–185.

107 Short, J. 1983. Density and seasonal movements of forest elephant (*Loxodonta africana cyclotis*, Matschie) in Bia National Park, Ghana. Afr. J. Ecol. 21: 175–184.

108 IUCN. 1987. Estimated elephant population sizes 1987. African Elephant and Rhino Specialist Group of the IUCN Species Survival Commission. (AERSG).

109 Roth, H. H., G. Merz et B. Steinhauer. 1984. Répartition et statut des grandes espèces de mammifères en Côte d'Ivoire, I. Introduction, II. Les éléphants. Mammalia 48/2: 207–226.

110 Merz, G. 1981. Recherches sur la biologie de nutrition et les habitats préférés de l'éléphant de fôret, *Loxodonta africana cyclotis*, Matschie 1900. Mammalia 45/3: 299–312.

111 Aubréville, A. 1958. A la recherche de la forêt en Côte d'Ivoire. Bois Forêt Trop. 57: 12–27.

112 Alexandre, D. Y. 1978. Le rôle disséminateur des éléphants en forêt de Taï, Côte d'Ivoire. La Terre et la Vie 32: 47–72.

113 Lieberman, D., M. Lieberman and C. Martin. 1987. Notes on seeds in elephant dung from Bia National Park, Ghana. Biotropica 19/4: 365–369.

114 Boesch, Christophe. 1988. pers. comm.

115 Schnelle, H. 1971. Die traditionelle Jagd Westafrikas. Diss. Univ. Göttingen.

116 Ayensu, E. S. 1978. Medicinal plants of West Africa. Reference Publications, Michigan. 330 pp.

117 Njikam, A. P. and C. N. Mbi. 1988. The role of the Centre for the Study of Medicinal Plants (CEPM) in the exploitation, development and production of drugs from plants with potential therapeutic value in tropical rain forests (Cameroon). Centre for the Study of Medicinal Plants MESRES, Yaoundé. Cyclostyled.

118 Ayensu, E. S. 1980. Jungles. Jonathan Cape Ltd., London.

119 Myers, N. 1985. Tropical moist forests: over-exploited and under-utilized? Report to WWF/IUCN. Gland, Switzerland.

120 Infield, M. 1988. Hunting, trapping and fishing in villages within and on the periphery of the Korup National Park. Paper No. 6, Korup N. P. socio-economic survey. WWF, Gland Switzerland.

121 Asibey, E. O. A. 1978. Wildlife production as a means of protein supply in West Africa, with particular reference to Ghana. Proc. of the VIII th World Forestry Congr. 869–885.

122 de Vos, A. 1977. Game as food. A report on its significance in Africa and Latin America. Unasylva 29: 2–12.

123 Ajayi, S. S. 1971. Wildlife as a source of protein in Nigeria: some priorities for development. The Nigerian Field, 36(3): 115–127.

124 Martin, G. H. G. 1983. Bushmeat in Nigeria as a natural resource with environmental implications. Environmental Conservation 10(2): 125–132.

125 Asibey, E. O. A. 1977. Expected effects of land-use patterns on future supplies of bushmeat in Africa south of the Sahara. Environmental Conservation 4(1): 43–50.

126 Asibey, E. O. A. 1974. Wildlife as a source of protein in Africa south of the Sahara. Biological Conservation 6(1): 32–39.

127 Asibey, E. O. A. 1974. Ecological and economic aspects of grasscutter in Ghana. Ph. D. thesis, University of Aberdeen, UK.

128 Orraca-Tetteh, R. 1963. The giant african snail as a source of food. In: The better use of the world's fauna for food. J. D. Ovington (ed.) 53–61. Institute of Biology, London.

129 Waitkuwait, W.E. 1987. Nutzungsmöglichkeiten der westafrikanischen Riesenschnecken (Achatinidae). Laboratoire Central de Nutrition Animal (LACENA), Abschlussbericht zur GTZ-Eigenmassnahme Nr. 85–9125.7–91.100.

130 Sale, J. B. 1981. The importance and values of wild plants and animals in Africa. IUCN, Gland, Switzerland.

131 Balinga, V. S. 1977. Competitive uses of wildlife. Unasylva 29: 22–25.

132 Léna, P. 1984. Le développement des activités humaines. Dans: Recherche et aménagement en milieu tropical humide, le projet Taï de Côte d'Ivoire (J. L. Guillaumet, G. Couturier et H. Dosso (eds.) Notes techniques du MAB 15, Unesco.

133 Asabere, P. K. 1987. Attempts at sustained yield management in the tropical high forests of Ghana. In: Natural management of tropical moist forests. F. Mergen and J. R. Vincent (eds.). Yale University.

134 FAO. 1985. Intensive multiple-use forest management in the tropics. Analysis of case studies from India, Africa, Latin America and the Caribbean. FAO Forestry Paper 55. FAO, Rome.

135 Nwoboshi, L. C. 1987. Regeneration success of natural management, enrichment planting and plantations of native species in West Africa. In: Natural Management of tropical moist forests. F. Mergen and J. R. Vincent (eds.). Yale University.

136 Kio, P. R. O. and S. A. Ekwebelan. 1987. Plantations versus natural forests for meeting Nigeria's wood needs. In: Natural management of tropical moist forests. F. Mergen and J. R. Vincent (eds.) Yale University.

137 FAO. 1986, 1989. Yearbooks of forest products, 1974–85, 1976–87. FAO, Rome.

138 Dorm-Adzobu, C. 1985. Forestry and forest industries in Liberia. An example in ecological destabilization. Int. Institute for Environment and Society, Berlin.

139 Page, J. M. 1978. Economies of scale, income distribution, and small-enterprise promotion in Ghana's timber industry. Food Research Studies XVI,3: 159–181.

140 Oldfield, S. 1988. Rare tropical timbers. IUCN Tropical Forest Programme. IUCN Gland, Switzerland and Cambridge, UK.

141 Nectoux, F. 1985. Timber! An investigation of the UK tropical timber industry. Friends of the Earth, London.

142 Nectoux, F. and N. Dudley. 1987. A hardwood story. An investigation into the european influence on tropical forest loss. Earth Resources Research Ltd., Friends of the Earth.

143 Westoby, J. 1978. Forest Industries for socio-economic development. Paper presented at World Forestry Congress, Jakarta.

144 Curry-Lindahl, K. 1969. Report to the government of Ghana on conservation, management and utilization of Ghana's wildlife resources. IUCN Morges, Switzerland.

145 Curry-Lindahl, K. 1969. Report to the government of Liberia on conservation, management and utilization of wildlife resources. IUCN Morges, Switzerland.

146 TFAP. 1985. The Tropical Forestry Action Plan. World Resources Institute, FAO, World Bank, United Nations Development Programme. FAO, Rome.

147 Budowski, G. 1984. The role of tropical forestry in conservation and rural development. Commission on Ecology Paper No. 7, Int. Union for Conservation of Nature (IUCN), Gland, Switzerland.

148 IUCN. 1985. United Nations list of national parks and protected areas. Int. Union for Conservation of Nature IUCN, Gland, Switzerland and Cambridge, U. K.

149 IUCN. 1986. Review of the protected areas system in the afrotropical realm. Compiled by J. and K. McKinnon, in collaboration with UNEP. IUCN, Gland, Switzerland.

150 IUCN. 1987. Directory of afro-tropical protected areas. IUCN, Gland, Switzerland and Cambridge U. K.

151 Rahm, U. 1973. Proposition pour la création du Parc national ivorien de Taï. Rapport d'une mission d'étude pour le compte de l'UICN et le WWF à l'intention du Ministère de l'Agriculture de Côte d'Ivoire. Morges, Suisse.

152 Asibey, E. O. A. and J. G. K. Owusu. 1982. The case for high-forest national parks in Ghana. Environmental Conservation 9(4): 293–304.

153 Verschuren, J. 1983. Conservation of tropical rainforest in Liberia. Recommendations for wildlife conservation and national parks, based on a survey of 1978. WWF/IUCN, Gland, Switzerland.

154 LRDC. 1987. Korup Project soil survey and land evaluation. Report 3206/8. WWF, Gland, Switzerland.

155 Devitt, P. 1989. Korup Project socio-economic survey. WWF, Gland, Switzerland.

Annex 1

Important West African timber species

Common trade names	Scientific name	Weight kg/m³
Abura (Bahia, Subaha)	Mitragyna ciliata	930
African greenheart (Sougue)	Parinari excelsa	1000
African mahogany (Khaya, Acajou)	Khaya ivorensis	750
Afrormosia (Kokrodua, Assamela)	Pericopsis elata	1200
Aningre (Aniegré blanc)	Aningeria robusta	1000
Antiaris (Ako)	Antiaris africana	550
Avodire (Wansenwa)	Turraeanthus africanus	750
Azobe (Bongossi, Red Ironwood)	Lophira alata	1300
Bosse (Krasse, Kwabohoro)	Guarea cedata	900
Canarium (Aiele, Abel, Labe)	Canarium schweinfurthii	750
Cardboard (Ilomba, Otie)	Pycnanthus angolensis	750
Dabema (Dahoma, Atui)	Piptadeniastrum africanum	960
Danta (Kotibe, Ahia)	Nesogordonia papaverifera	960
Doussie (Lingue, Apa)	Afzelia spp.	1100
Emeri (Framire, Deohr, Idigbo)	Terminalia ivorensis	780
Gedu nohor (Tiama, Edinam)	Entandrophragma angolense	830
Iroko (African oak, Odum)	Chlorophora excelsa	950
Khaya I (Acajou blanc, Ahafo)	Khaya anthotheca	750
Khaya II (Acajou à grandes feuilles)	Khaya grandifolia	750
Kosipo (Heavy Sapele, Penkwa)	Entandrophragma candollei	950
Limba (Frake, Ofram)	Terminalia superba	800
Lovoa (Dibetou, African Walnut)	Lovoa trichilioides	750
Makore (Baku, Abako)	Tieghemella heckelii	860
Mansonia (Bete, Aprono)	Mansonia altissima	930
Movingui (Ayan, Distemonanthus)	Distemonanthus benthamianus	950
Niangon (Angi, Ogoue)	Tarrietia utilis	910
Obeche (Wawa, Abachi, Samba)	Triplochiton scleroxylon	560
Ogea (Faro, Gum Copal)	Daniella spp.	800
Okwen (Akume, Meblo)	Brachystegia spp.	850
Opepe (Bilinga, Badi)	Nauclea diderichii	1050
Ovengkol (Amazakoue, Hyedua)	Guibourtia ehie	1050
Pterygota (Koto, Kefe)	Pterygota macrocarpa	900
Sapele (Sapelli, Aboudikro, Penkwa)	Entandrophragma cylindricum	870
Sasswood (Tali, Missanda)	Erythrophleum ivorense	1150
Sikon (Ekaba, Hoh)	Tetraberlinia tubmaniana	900
Utile (Sipo, Assie)	Entandrophragma utile	780

The list only includes the most common trade names. Numerous local names and, in the case of Central African destinations, other trade names are also in use (see Dahms, 29). The weight indications refer to freshly harvested roundwood. The dry weights per m³ are 30–40 % less.

Annex 2

Mammals occurring in Taï National Park (Côte d'Ivoire), according to Roth and Merz [71]

	Number of species per family
Insect-eaters (*Insectivora*)	
Shrews (*Soricidae*)	14
Bats (*Chiroptera*)	
Fruit bats (*Pteropodidae*)	12
Sac-winged bats (*Emballonuridae*)	1
Slit-faced bats (*Nycteridae*)	4
Horseshoe-bats (*Rhinolophidae*)	3
Old world leaf-nosed bats (*Hipposideridae*)	7
Vespertilionid bats (*Vespertilionidae*)	9
Free-tailed bats (*Molossidae*)	7
Rodents (*Rodentia*)	
Rats and mice (*Muridae*)	21
Dormice (*Gliridae*)	3
Flying squirrels (*Anomaluridae*)	4
– Fraser's flying squirrel (*Anomalurus derbianus*)	
– Pel's flying squirrel (*Anomalurus peli*)	
– Beecroft's flying squirrel (*Anomalurus beecrofti*)	
– Pygmy flying squirrel (*Idiurus macrotis*)	
Cane rats (*Thryonomyidae*)	1
– Large cane rat (*Thryonomys swinderianus*)	
Squirrels (*Sciuridae*)	8
– Striped ground squirrel (*Euxerus erythropus*)	
– Giant forest squirrel (*Protoxerus stangeri*)	
– Slender-tailed squirrel (*Protoxerus aubinni*)	
– Red-legged sun squirrel (*Heliosciurus rufobrachium*)	
– Palm squirrel (*Epixerus ebii*)	
– Small green squirrel (*Paraxerus poensis*)	
– Red-footed squirrel (*Funisciurus pyrrhopus*)	
– Green side-striped squirrel (*Funisciurus substriatus*)	
Porcupines (*Hystricidae*)	2
– Brush-tailed porcupine (*Atherurus africanus*)	
– Crested porcupine (*Hystrix cristata*)	
Giant rats (*Cricetidae*)	2
– Emin's giant rat (*Cricetomys emini*)	
– Gambian giant rat (*Cricetomys gambianus*)	

	Number of species per family
Pangolins (*Pholidota*)	
Pangolins (*Manidae*)	3
– Tree pangolin (*Manis tricuspis*)	
– Long-tailed pangolin (*Manis tetradactyla*)	
– Giant pangolin (*Manis gigantea*)	
Primates (*Primates*)	
Loris (*Lorisidae*)	1
– Bosman's potto (*Perodicticus potto*)	
Galagos (*Galagidae*)	1
– Dwarf galago (*Galago demidovii*)	
Cercopithecid monkeys (*Cercopithecidae*)	5
– Campbell's mona monkey (*Cercopithecus campbelli*)	
– Diana monkey (*Cercopithecus diana*)	
– Lesser spot-nosed monkey (*Cercopithecus petaurista*)	
– Greater spot-nosed monkey (*Cercopithecus nictitans*)	
– Sooty mangabey (*Cercocebus atys*)	
Colobus monkeys (*Colobidae*)	3
– Olive colobus (*Colobus verus*)	
– Red colobus (*Colobus badius*)	
– Black-and-white colobus (*Colobus polykomos*)	
Great apes (*Pongidae*)	1
– Chimpanzee (*Pan troglodytes*)	
Carnivores (*Carnivora*)	
Mustelids (*Mustelidae*)	3
– Cape clawless otter (*Aonyx capensis*)	
– Spotted-necked otter (*Lutra maculicollis*)	
– Ratel (*Mellivora capensis*)	
Viverrids (*Viverridae*)	6
– Marsh mongoose (*Atilax paludinosus*)	
– Dark mongoose (*Crossarchus obscurus*)	
– Liberian mongoose (*Liberiictis kuhni*)	
– African civet (*Viverra civetta*)	
– Two-spotted palm civet (*Nandinia binotata*)	
– Large-spotted genet (*Genetta tigrina*)	
Cats (*Felidae*)	2
– Golden cat (*Profelis aurata*)	
– Leopard (*Panthera pardus*)	

	Number of species per family
Proboscideans (Proboscidea)	
Elephants (Elephantidae)	
– African elephant (Loxodonta africana)	1
Hyraxes (Hyracoidea)	
Hyraxes (Procaviidae)	
– Tree hyrax (Dendrohyrax arboreus)	1
Even-toed ungulates (Artiodactyla)	
Wild pigs (Suidae)	
– Giant forest hog (Hylochoerus meinertzhageni)	
– Bush-pig (Potamochoerus porcus)	2
Hippopotamus (Hippopotamidae)	
– Pygmy hippopotamus (Choeropsis liberiensis)	1

	Number of species per family
Chevrotains (Tragulidae)	
– Water chevrotain (Hyemoschus aquaticus)	1
Horned ungulates (Bovidae)	
– Blue duiker (Cephalophus monticola)	
– Black duiker (Cephalophus niger)	
– Bay duiker (Cephalophus dorsalis)	
– Banded duiker (Cephalophus zebra)	
– Ogilby's duiker (Cephalophus ogilbyi)	
– Yellow-backed duiker (Cephalophus silvicultor)	
– Jentink's duiker (Cephalophus jentinki)	
– Royal antelope (Neotragus pygmaeus)	
– Bushbuck (Tragelaphus scriptus)	
– Bongo (Boocerus euryceros)	
– African buffalo (Syncerus caffer)	11

In addition to the 140 species recorded, the occurrence of the following species is possible or likely: 6 further species of bats, 2 species of genet cats (Genetta pardina and G. johnstoni) and the African linsang (Poiana richardsoni). Only the most common English and scientific names are listed here. Local subspecies or forest forms (e.g. forest elephant Loxodonta africana cyclotis) are not distinguished.

Index of Authors Cited

The numbers in brackets refer to the publications cited in the text and under References on page 219 ff.

Subject Index

Index of Photos and Figures

Thanks

Dr. Emmanuel O.A.Asibey acquainted me with many aspects of the rainforest during my years in Ghana. John B. Hall, who has since passed away, gave me an insider's view into forest botany. I owe them both endless gratitude. I would also like to thank Dr. Peter Duelli of the Swiss Institute for Forest, Snow and Landscape Research for his comments on the zoological chapters and Chris Elliott, Senior Forest Conservation Officer of WWF International for his editing assistance. My wife Judy worked on the original manuscript and her untiring support played a great role in the writing of this book. I am also very grateful for the cooperation of Andreas Bally, Albert Gomm and Irina Weiss of Birkhäuser Publishers.

C. M.